HUMANISTIC MEDICAL INTERIORS IN CHINA

中国人文医疗空间开拓者

亚明大型医院室内设计

Yaming Interior Design Company Ltd.

孙亚明 / 编著 孙哲 / 译

辽宁科学技术出版社

ABOUT THE AUTHOR
关于作者

Sun Yaming

Senior Interior Desiner
China Top 10 Hospital Interior Designer
Tsinghua University Architecture Engineering & Design Advanced Class
Interior Design Master of Polytechnic University of Milan
Vice Chairman of the Nanjing Institute of Interior Design
Distinguished Professor of Artistic Design Department of Nanjing University Jinling College
Executive Director of Jiangsu Interior Design Association
Standing Committee of Hospital Architecture & Equipment Chapter of China Association of Medical Equipment

Yaming Interior Design Company Ltd. has won the Luban Prize, the highest rank for construction project in China, for five times. Other prizes and awards are listed as follows:
2006, the first prize of Army Design Competition
2005, the first prize of Second Cross-straits Interior Design Competition
2005, Golden, Silver, and Bronze awards of 4th Jiangsu Interior Decorative Design Contest
2012, Golden Award of Asia-Pacific Hotel Association Design Contest
2014, the first prize of Nanjing Interior Design Competition
By now, Yaming Interior Design Company Ltd. has completed nearly 300 hospital designs and it currently ranks 1st in national medical design filed.

孙亚明

高级室内建筑师
中国十佳医院室内设计师
清华大学建筑工程与设计高级研修班
意大利米兰理工大学室内设计硕士
南京市室内设计学会副会长
南京大学金陵学院艺术设计系特聘教授
江苏省室内装饰协会常务理事
中国医学装备协会医院建筑与装备分会常务委员

亚明设计曾五次获得中国建筑行业最高奖——鲁班奖，2006 年全军设计一等奖，2005 第二届海峡两岸四地室内设计大赛一等奖，并囊括江苏省第四届室内装饰设计大奖赛金奖、银奖、铜奖，2012 亚太酒店设计协会大奖赛金奖，2014 南京市室内设计大赛一等奖。在国内已经设计了近 300 家医院，目前在国内医疗设计领域业绩排名第一。

17 年的专注，

近 300 家医院落地了亚明的设计

孙亚明

PREFACE
序言

Seventeen years ago, I found Nanjing the cradle of my dream. Since then, like a hermit traveling thousands of miles barely on foot to obtain true Buddhism sutra, I led a simple and humble life, marching forward day and night regardless of hardship to fulfill my dream. Thanks to the people whose efforts and wisdom contributed to the growth of Yaming Interior Design Company Ltd., it has already developed into a larger scale containing dozens of avant-garde designers. Ever since the millennium, as many as 300 hospitals successfully implemented our design proposals, many of which have won awards both at home and abroad.

This is the full recognition by our clients and the huge reward to our brand. On this occasion, I sincerely express my gratitude to my colleagues who spent day and night helping me overcome all obstacles. Also, I am extremely grateful to my family for their understanding and support.

The widely accepted principle that scenery is the product of the nature has been followed and adhered to in most hospital design projects. And we consistently set the prime goal of medical function solution and medical procedure improvement in our hospital designs, aiming to supply the patients and the medical personnel with more scientific, reasonable and humanized design strategies. In the meanwhile, we keep constant innovation to convey humanistic concerns to the mass of people in the medical environment. Though the road of design is like an endless marathon, we, united as a vigorous team, are ready to make persistent efforts to reach a higher level. We are confident that with our ingenuity, more interior designs of high quality will be dedicated to our clients in the near future.

By the Qinhuai Riverside
In the Spring, 2017

亲爱的朋友们：

17 年前，梦想的征途开启，我将这条路的起跑点设在了南京。为了实现心之所望，遂素履以往，披星戴月，一路风雨兼程，却也甘之若饴。将公司从当初的几人发展至如今的几十人，前行的每一步，都有着对公司发展的殚精与竭虑。千禧年开始至今，已经有近 300 家医院落地了我的设计，这些作品也斩获了海内外多项设计大奖。这是大家对亚明工作的最大肯定，也是对无数个工作加班不眠夜最好的褒奖。感谢这个时代，它让我有不断努力的激情；感谢我的业主，他们给了我进步的机会；也感谢我的妻儿，他们让我有了面对困难的动力；同时也要感恩我的同事，他们让我变得更有责任感。

犹记得入行之初，在医院实地考察门诊、医技到护士站、处置治疗室以及病房等功能空间，请教每一位医生的日夜。在不断的总结与前行中，坚持将解决医疗空间的功能和医疗工艺流程系统作为我设计的首要目标，致力于为医护以及病患提供更科学合理、更具人性化的空间设计。在服务客户的同时，亚明也在不断创新，研发出的工业系统装配集成施工体系已经在众多项目中运用，将极简的外在与丰富的人文关怀传达给每一位身处在医疗空间的芸芸众生。

“不忘初心，方得始终”，相信梦想的力量，吾辈当与之同行。十七年，再次感谢一路上有你！

孙亚明
二零一七年阳春于金陵秦淮河畔

CONTENTS
目录

江苏省人民医院

Jiangsu Province Hospital

始于 1936 年的江苏省人民医院，暨南京医科大学第一附属医院。目前是江苏省综合实力最强的三级甲等医院。将南京的市花梅花、医学分子、吉特达木人元素融入到设计中。使得医院空间更具人文关怀。医院内部空间以极少的材料，构建出明亮友好的色调。清晰的造型手法为人们营造了一种开放、吸引人的氛围。医患人员身处其中，有助于增进互信，就诊人员置身于不同以往的就医环境中，能够更快地调节自己的焦虑心情，配合医护人员进行诊疗。

Founded in 1936, Jiangsu Province Hospital, the first affiliated hospital to Nanjing Medical College, is contemporarily the greatest comprehensive class-A hospital in Jiangsu province. The integrated design elements of plum blossom (the city-flower of Nanjing), medical molecules and Target wooden man increase its humanistic concern. The minimum use of materials in the interior space creates a friendly bright tone, and the technique for concrete modeling builds an open and appealing atmosphere, which is likely to enhance the mutual trust between patients and doctors. Standing in this distinctiove medical environment, the patients can easily adjust their anxiety and cooperatively receive treatment.

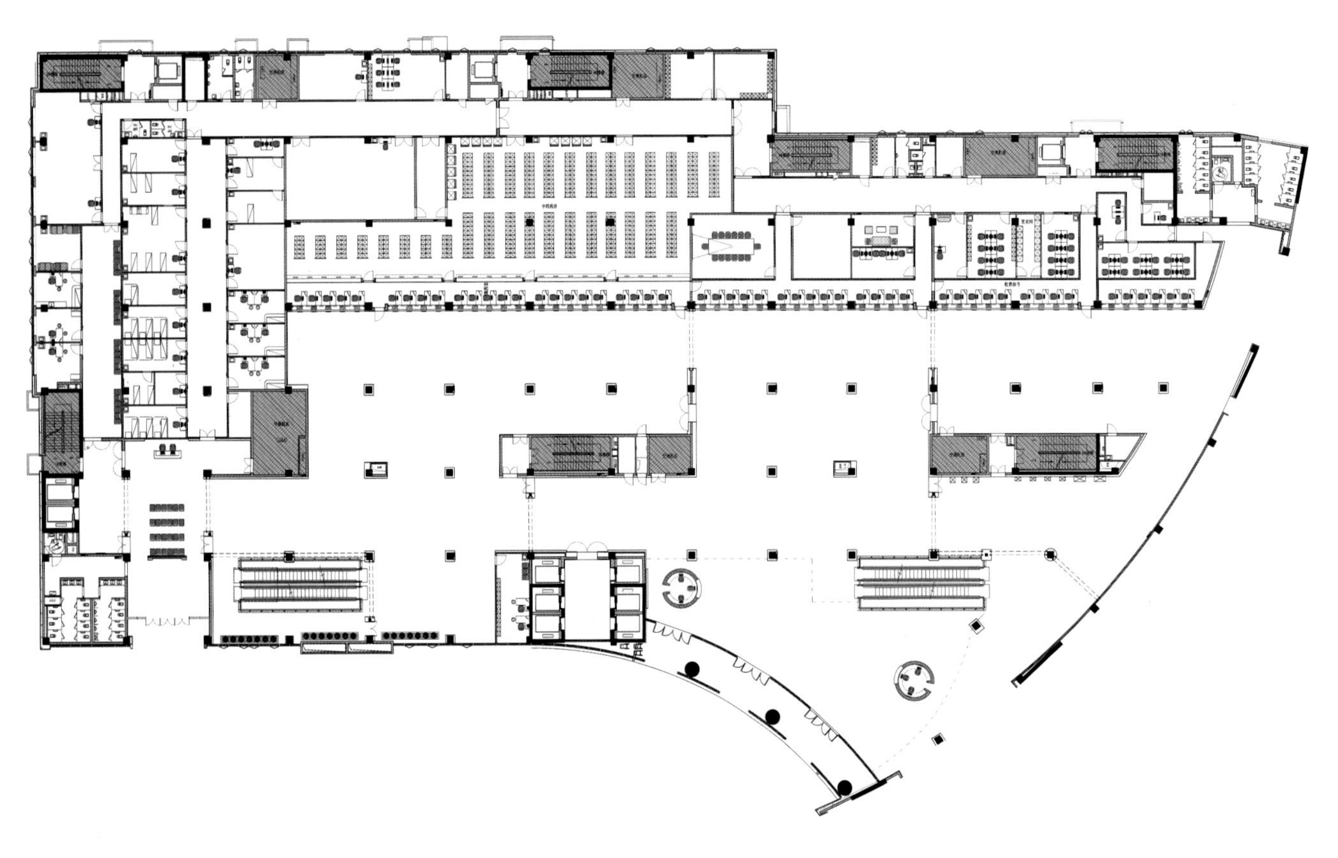

2F

护士站 | NURSE STATION
NURSING
STAITON

internet

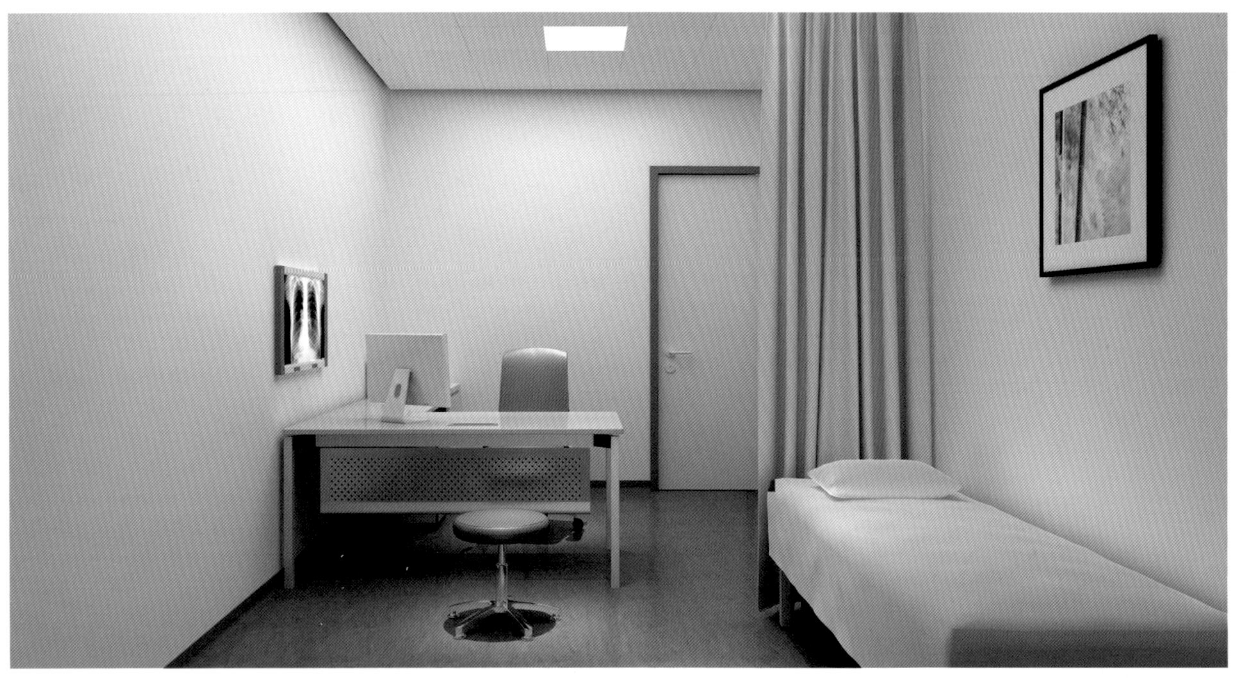

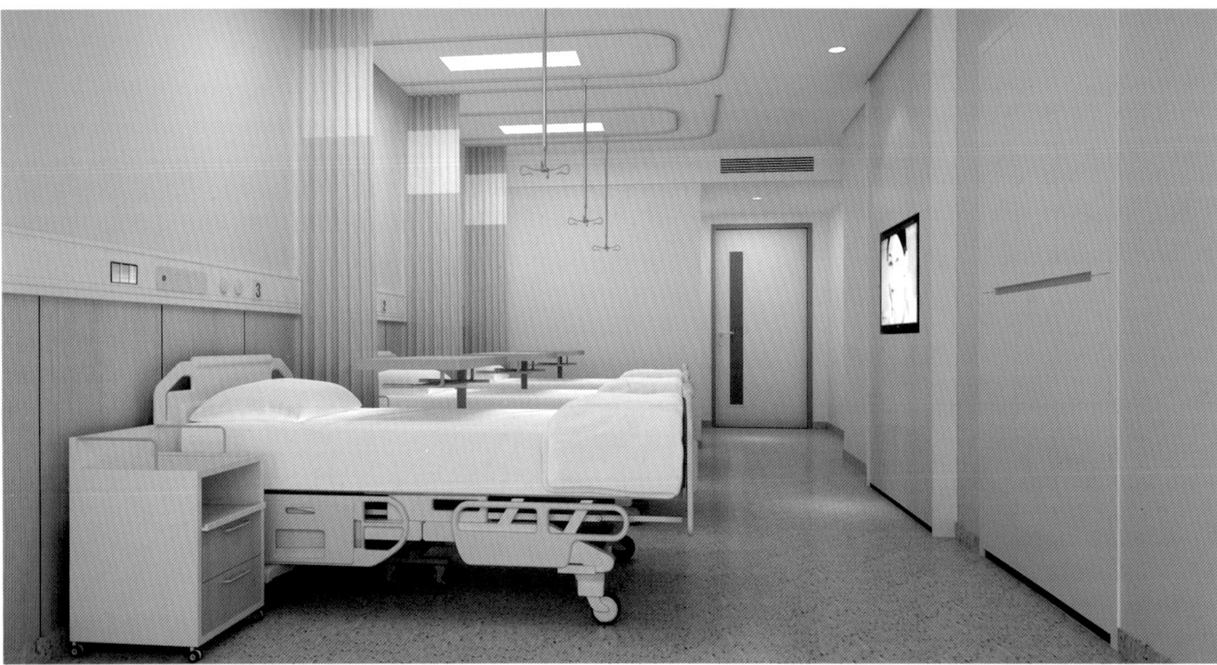

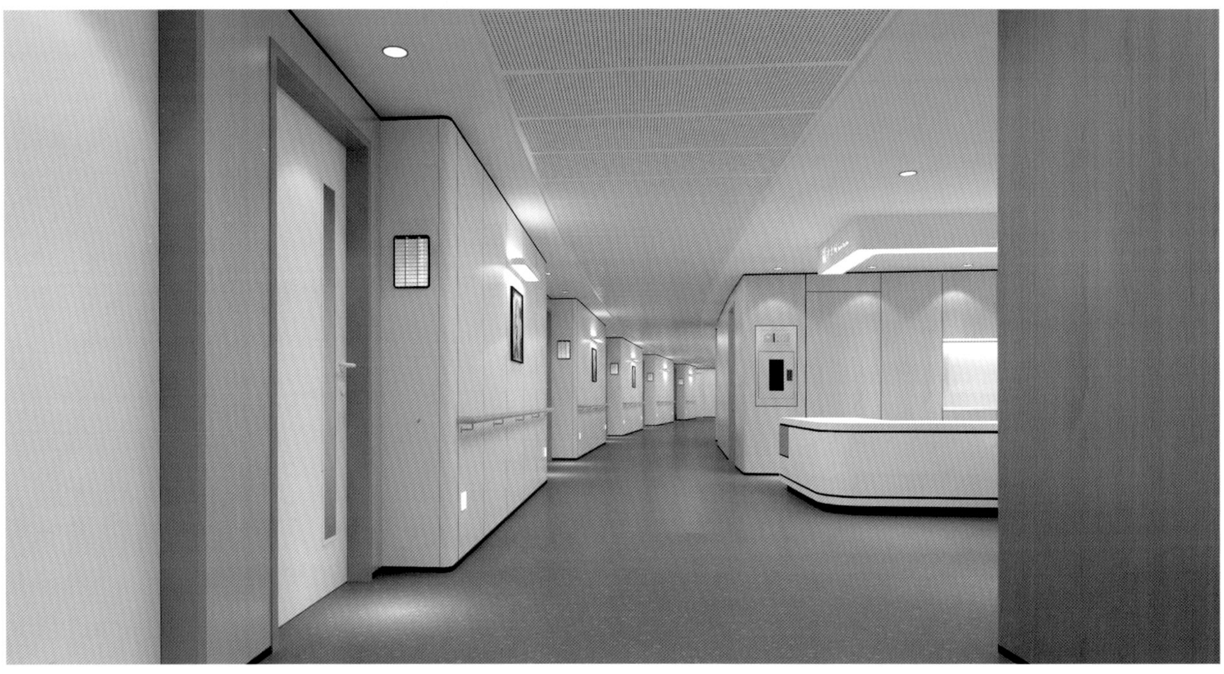

出于对医生使用习惯的考虑，诊桌设计为右手靠墙。在保护病人隐私上，L 形隔帘优于普通类型隔帘。护士站采用圆弧设计，避免对人造成伤害的同时释放温和的空间氛围。顶部采用可拆卸材料，避免混响的同时，更便于管线维护与升级，走道采用泛光照明方式，避免灯光直射人眼。

Considering the habits of doctors, all clinical stages in the hospital are right-handed against the wall. In terms of the patient privacy protection, L-shaped curtains are superior to the normal ones. The nurse station uses the arc design to avoid harm to the human beings and also creates a mild atmosphere. The demountable ceiling can be used to avoid the reverberation at the same time and the pipelines are easier to maintain and upgrade. The corridor uses a flood lighting to avoid direct light to eyes.

水幕墙及花草设置增加了空间的灵动之气，艺术品隔断巧妙地将空间与走廊区分开来。

The arrangement of water retaining walls and plants add flexibility to the space, and the fine arts partitions tactfully separate clinic and corridor areas.

通道区域的墙面采用新型材质使得墙面更具肌理感，增设院史内嵌展览橱窗意在传达医院的历史底蕴。

New materials are employed in the construction of aisle walls, which add more texture when touching. The embedded window exhibition of the hospital history intends to convey its inherent connotations.

SIPING CENTRAL HOSPITAL
2013
2014
SIPING CENTRAL HOSPITAL

医学开拓健康研讨会

江苏省人民医院GCP

Jiangsu Province Hospital GCP

江苏省人民医院 GCP 与国际医学团队有着良好的合作关系，在设计中使用梧桐元素体现南京钟灵坡秀的人文情节，同时将水立方的多边形外观作为自然元素应用其中。为了改变 GCP 过于“冰冷”的既有印象，设计以国际化、简洁化、科技感、未来感体现。使得空间整体简而不单，区别于既有的临床试药科研空间。

江苏省人民医院 GCP，建筑面积 2000m^2，以创新保质的系统集成设计缩短建设周期，在色彩搭配上米白色与灰色为主色，将色彩运用控制在三种以内，使得空间避免色彩杂乱而带来视觉混乱。

Jiangsu Province Hospital GCP has good cooperative relations to many international medical teams. Phoenix trees, the presentation of nature bestows upon the city and the polygon appearance of the Water Cube as the natural elements are integrated in its design. For the sake of changing the icy-cold impression of GCP, the design presents itself in an international, simplified, technological and prospective way. Therefore, the overall space is simple but not monotonous, and is distinctive from usual clinical medicine research area.

The construction period of Jiangsu Province Hospital GCP with a building area of 2,000 square meters is cut by innovative, quality-ensured, and integrated design system. In color matching, the architecture features creamy white and gray, and the choice of its colors is limited within three, which shields the space from color and visual clutter.

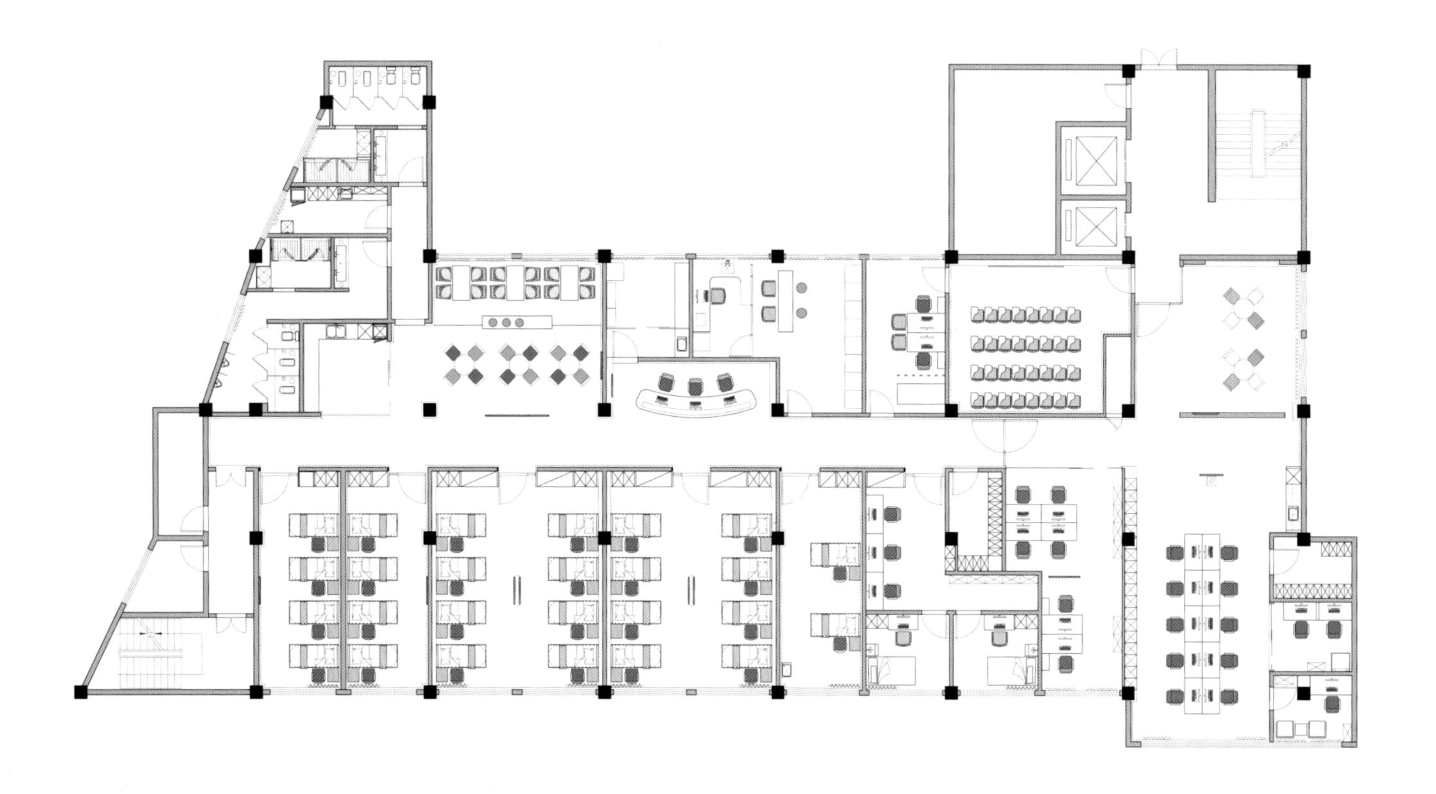

服务台采用更柔和的圆弧设计，台身的原木色成为灰白色空间中的点睛之笔，便于医患迅速识别，及时互动。

The circular arc design is adopted in the reception desk, with its tailored log color as the best line, is convenient for the patients' prompt recognition and immediate interaction.

NURSE STATION
江苏省人民医院

办公空间顶部穿孔石膏板材料可吸收嘈杂音，墙面元素与空间工作性质相联系，增加工作人员代入感。

The perforated plasterboard on top of the office area can absorb noise. The decorative elements on the wall have relation to the research work, which enforce the sense of identity for the staff.

江苏省南京鼓楼医院仙林国际医院

Nanjing Xianlin Drum Tower Hospital

南京鼓楼医院仙林国际医院坐落于古都南京城东的仙林地区，周边群山环绕，是紫金山东麓生态环境最完好的区域之一。设计时秉持将自然景观与艺术表现相结合即是美的初衷，以江水、青竹、分子元素彰显“以人文本，尊重生命”的设计理念。用最简单，最自然的设计语言对话空间。其中以水元素为概念的大厅灯具以及环绕支柱的等候休息座椅成为步入医院后最吸引人眼球的亮点。

Nanjing Xianlin Drum Tower Hospital is located in Xianlin area of eastern Nanjing, one of the best ecological environment regions of Zijin Mountain eastern foothills. Adhering to the premier principle that beauty lies in sceneries and artistic expressions, the design manifests its philosophy–human-orientation and respect for life, through the elements of mountains, bamboos and molecules. By the most natural and simplest words to communicate with the space, the lamps and the resting chairs embracing pillars become the spotlights at the entrance of the hospital.

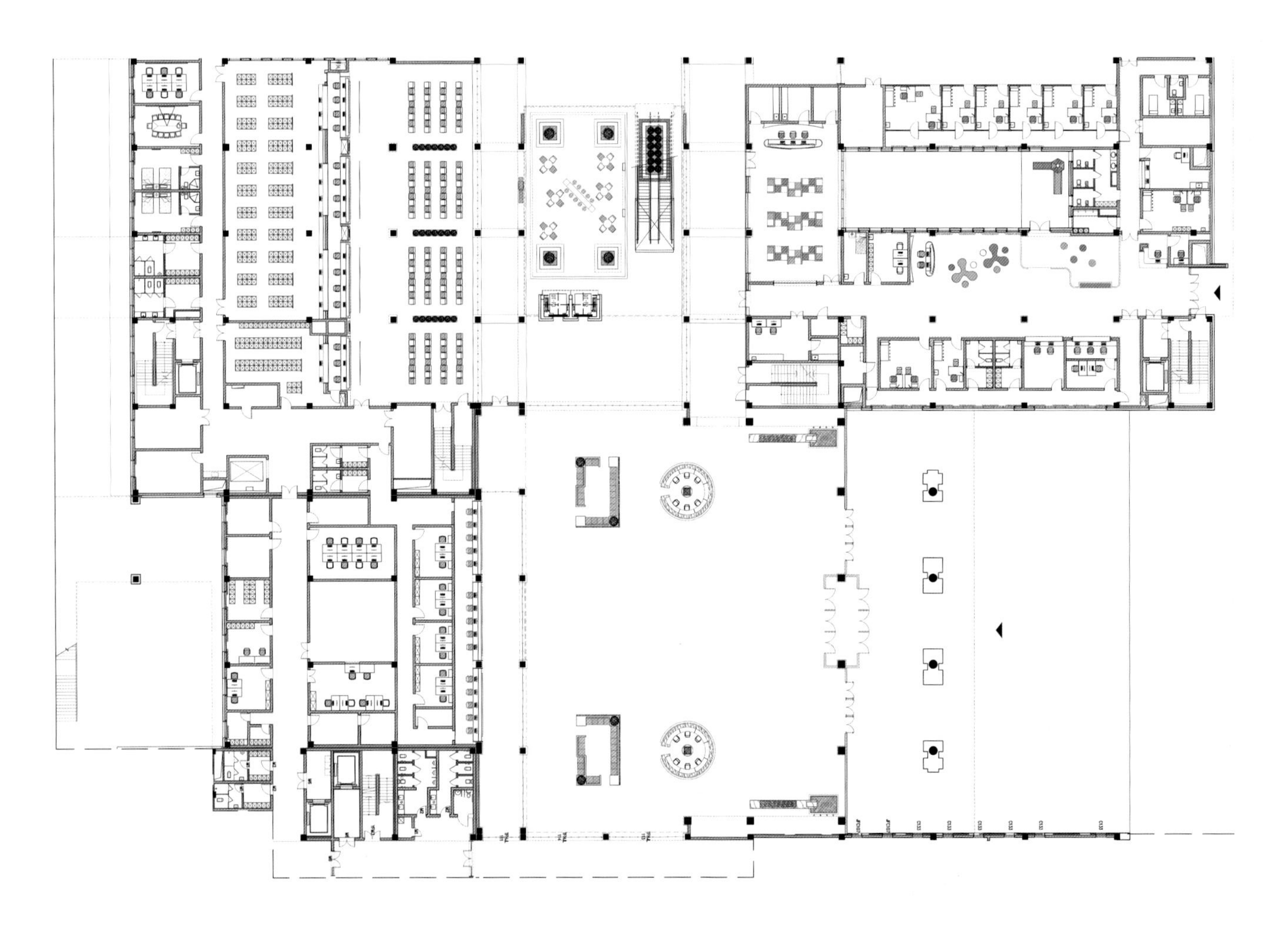

南京鼓楼医院仙林国际医院
XIANLIN INTERNATIONAL HOSPITAL

COFFEE
收费
收费
收费
等候区
WAITING AREA

中庭休闲区种植绿植，将自然在医院的室内呈现，给病患以积极向上的心理暗示。

Green plants are planted in leisure areas on the street, which presents nature and offers the patients positive psychological hints as well.

诊室

检验科
LABORATORY

1F

南京鼓楼医院仙林国际医院，建筑面积 150000m²，目前拥有床位 600 张。设计中通过材料以及颜色的运用，使得人在视觉、嗅觉、听觉、味觉、触觉上增进体验感，绿色亦成为该医院最具生命力的象征。

Nanjing Xianlin Drum Tower Hospital with a building area of 150,000 square meters opens 600 beds at present. By selected colors and materials, the design upgrades the patients' experience in five senses. The color of green is the most symbol of vitality in the hospital.

2
2
Area Index

护士站
Nurse Station
3F
病房 5

江苏省南京河西儿童医院

Nanjing Children's Hospital

南京河西儿童医院为原广州路儿童医院新院址，是一所集医疗、科研、康复、保健为一体的三级甲等儿童医院。本设计的难点在于，让儿童病患免于对医院空间的抵触感。在设计中大胆将奔放草原、奇幻海底、神秘太空作为设计元素。通过多元的色彩表达、生动的卡通形象、装饰用材的变化以及儿童游戏空间的引入等方式，将一所更具童真的医院呈现给大众。儿童病患在就医期间可以被空间内多样化的表现元素所吸引．从而提高医院的就诊效率。

As the new campus of Guangzhou Road Children's Hospital, Nanjing Children's Hospital is a comprehensive class-A hospital with the integration of medical care, scientific research, rehabilitation and health care. The difficulty of design is how to eliminate children's mental conflict with hospitals and therefore, unrestrained grassland, fantastic sea world and the mysterious space are boldly applied. Through a variety of colors, lively cartoon images, decorative material changes and game zones for children, a hospital possessing child's innocence is presented to the public. Children are so likely to be attracted by the varieties of expressing factors that their clinic efficiency is promoted during the period of medical treatment.

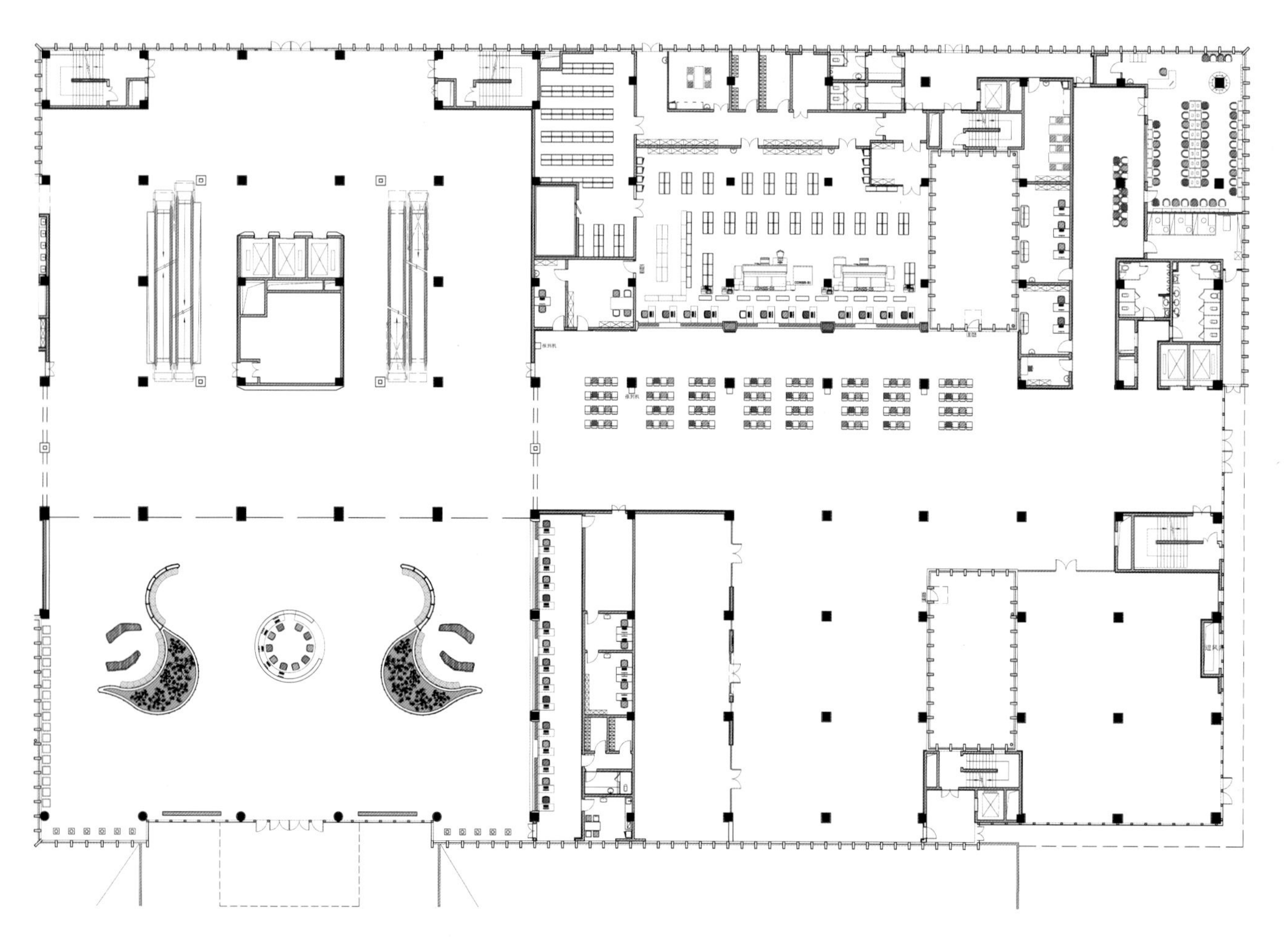

2F
挂号收费
Registration Fees
取药窗口
Dispensary Window

SUMMER
2015
NURSE STATION

用心呵护儿童健康

为了便于儿童以及残疾人使用，取药窗口的服务台采用高低台的设置。将液晶屏替代传统 LED 屏幕，避免与儿童氛围的空间相冲突。

For the convenience of children and the handicapped, the Pharmacy service counters adopt double-level design. The traditional LED screen is also replaced with LCD screen so that the conflict between the screens and the atmosphere is eliminated.

电梯以及自动扶梯均用色彩装饰，并有卡通图案点缀，使得空间中充满活跃气息。

Both elevators and escalators are colored packaged, and are embellished with cartoon pictures which make the space alive with vitality.

2F
挂号收費
取药窗口
六病區
二病區

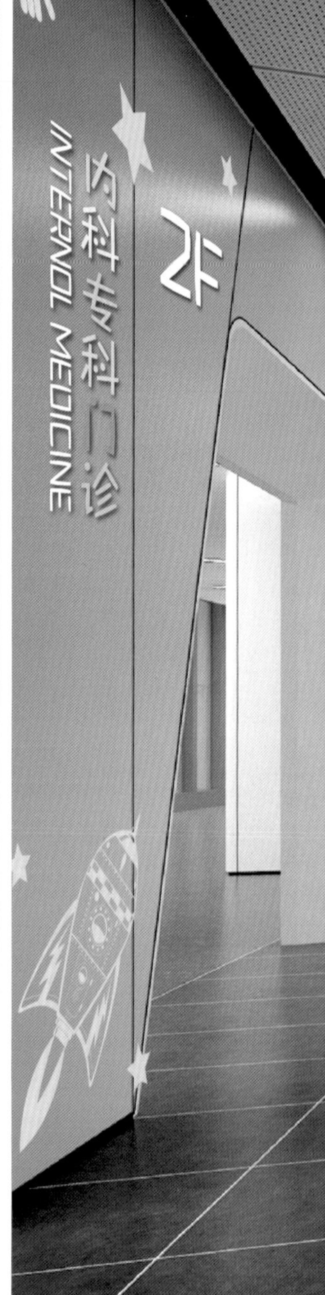
2F
内科专科门诊
INTERNOL MEDICINE

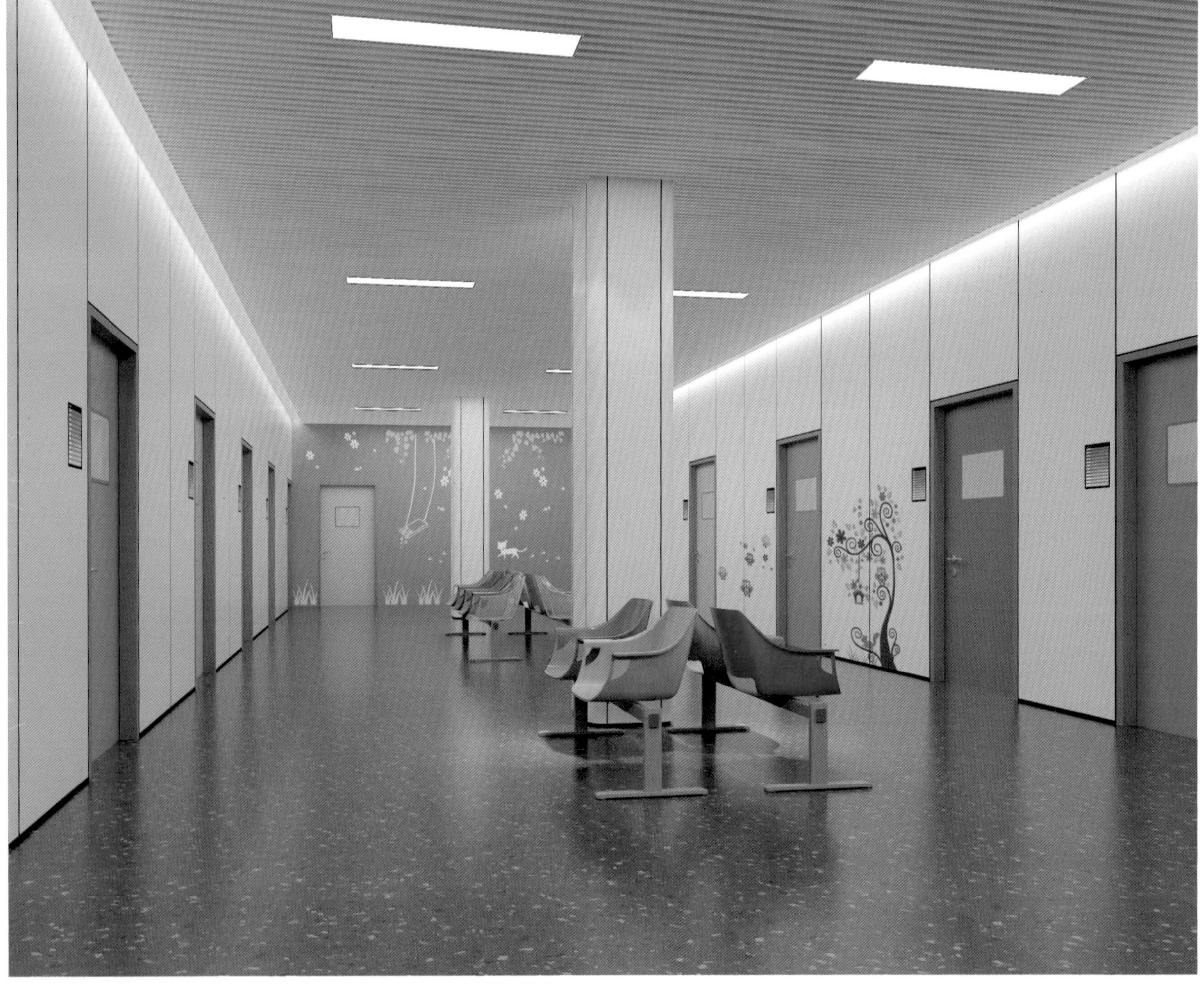

南京河西儿童医院，建筑面积 168000m^2，拥有床位 1100 张。既活泼又不失秩序的设计，使得该医院的主要群体：家长与儿童能够有自我意识的调整对陌生环境的紧张情绪，并能及时缓解家长的焦虑与儿童的畏惧感。

Nanjing Children's Hospital with a building area of 168,000 square meters opens 1,100 beds. The lively and orderly design enables parents and children, the major visiting group of the hospital, to intentionally adjust their tension and anxiety toward the new environment in the hospital.

NURSE STATION
NURSE STATION

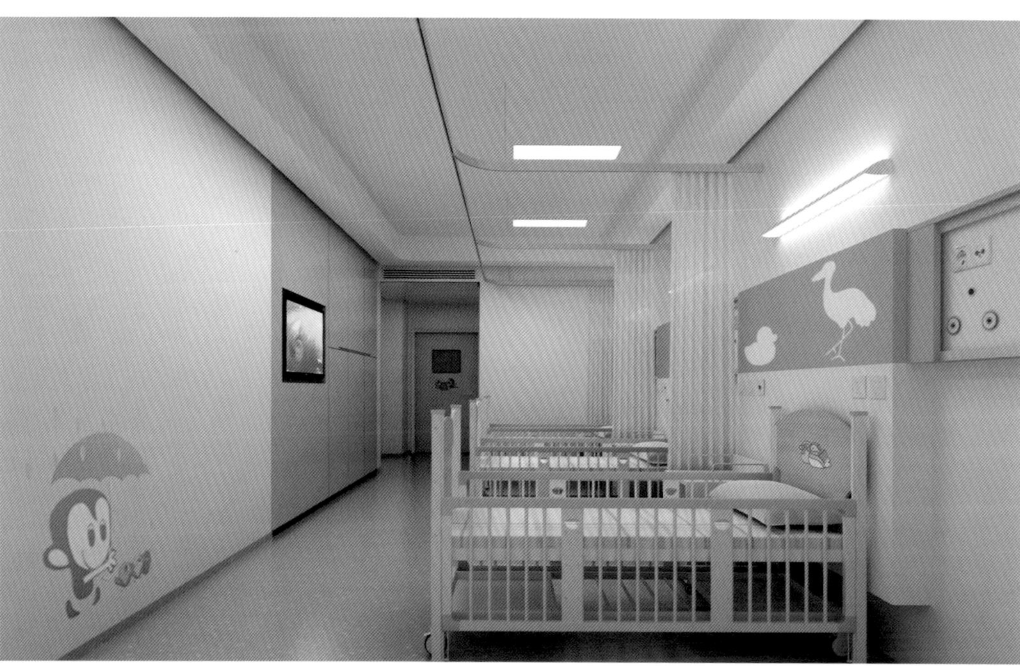

江苏省南京市第一医院

Nanjing First Hospital

南京市第一医院是所三级甲等综合性医院，又是南京医科大学附属医院之一，同时还是南京医科大学第三临床医学院、南京市红十字第一医院。地处夫子庙区域，是南京医科大学附属医院之一。设计将老城南的粉墙瓦黛，梅花与梧桐元素融入其中，在形式上强调精简洗练，营造“闲梦江南梅熟日”之感。

As a comprehensive class-A hospital, Nanjing First Hospital, known as No.3 Clinical Medicine College of Nanjing Medical University and Nanjing Red Cross Hospital, is located in the neighborhood of the Confucius Temple. It is also an affiliated hospital of Nanjing Medical University. The design stresses simplification in form with the integrated elements of pink walls, dark green roofs, plum blossoms and phoenix trees, and creates an experience of leisure life in Jiangnan towns.

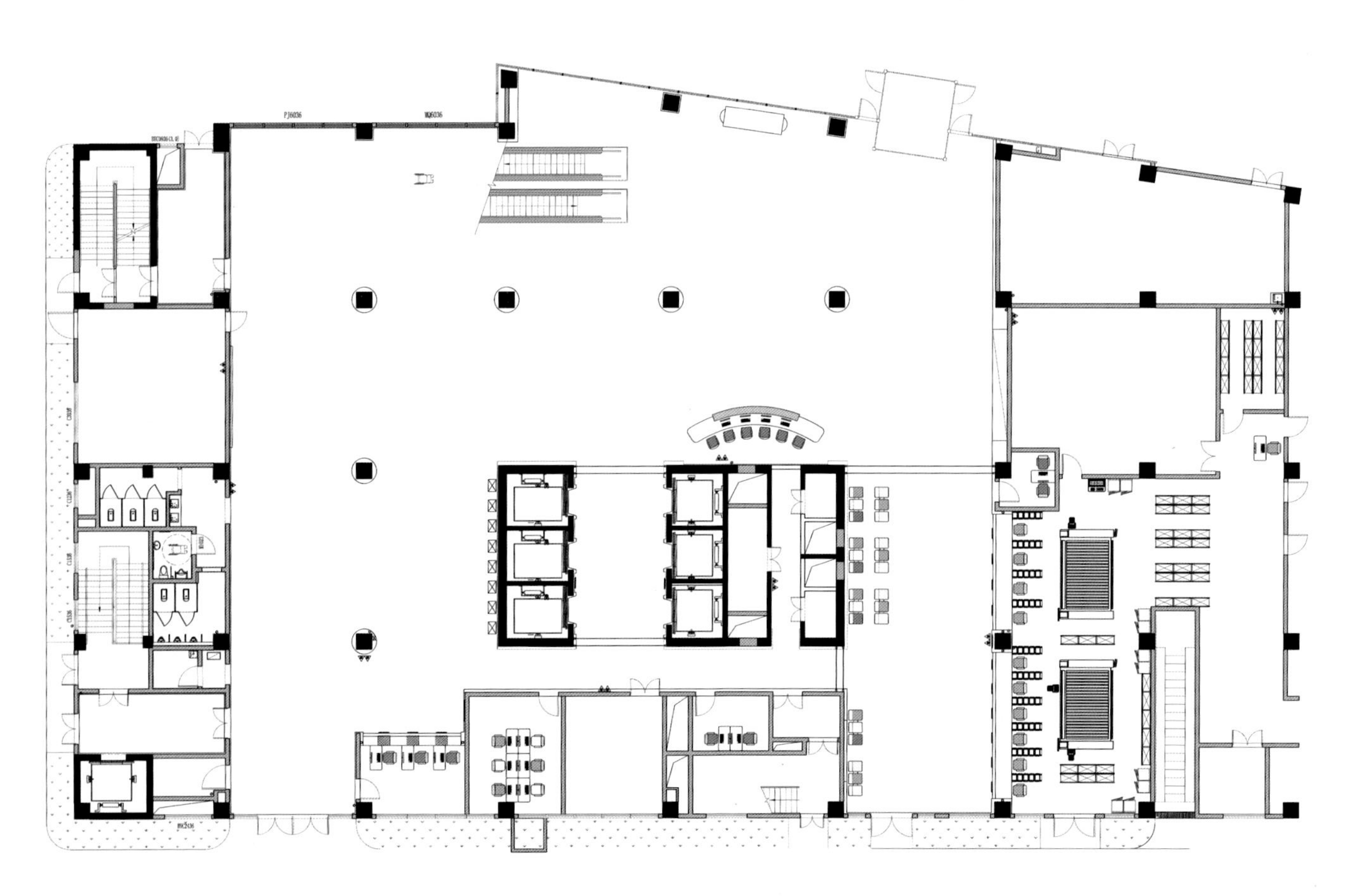

NURAE STATION

2▼
▼2

婴儿打理台

Windows 7

Windows 8

江苏省口腔医院（南医大附属口腔医院）

Jiangsu Province Stomatological Hospital

江苏省口腔医院暨南京医科大学附属口腔医院，是江苏省第一所三级甲等口腔专科医院。《汉书》有云，“目若悬珠，齿若编贝。”在南京医科大学附属口腔医院的设计中，微笑、钻石、医学分子等与牙齿相关的元素被植入到空间中。即使人们愿意去看牙医，但是钻头声音响起时到底是有所恐慌的，设计中除了在顶部设置电视机转移患者注意力外，严谨、亲切的几何护士站、支柱等都给人以钝性的印象，不会联想到牙痛。空间用简洁有力的材料说明自然、科技、健康的纯粹，给病患以镇静、安全感。

Jiangsu Province Stomatological Hospital, the affiliated hospital of Nanjing Medical College, is the first class-A stomatological hospital in Jiangsu province. As the quotation from *Book of Han* says "As one's bright eyes are shining like pearls, one's beautiful teeth are white as pearl buttons", the related elements like smile expression, diamond molecules, etc are implanted in the space design. People are frightened upon hearing the drill, even if they are ready to see the dentist. Therefore, apart from the TV set in the ceiling, the nurse station and the pillars, rigorous but kind, also leave people the impression of bluntness rather than sharpness, usually associated with the pain feeling as they are receiving treatment. Less but more powerful materials in the space demonstrate the purity of nature, health, science and technology, which gives patients a sense of tranquility and safety.

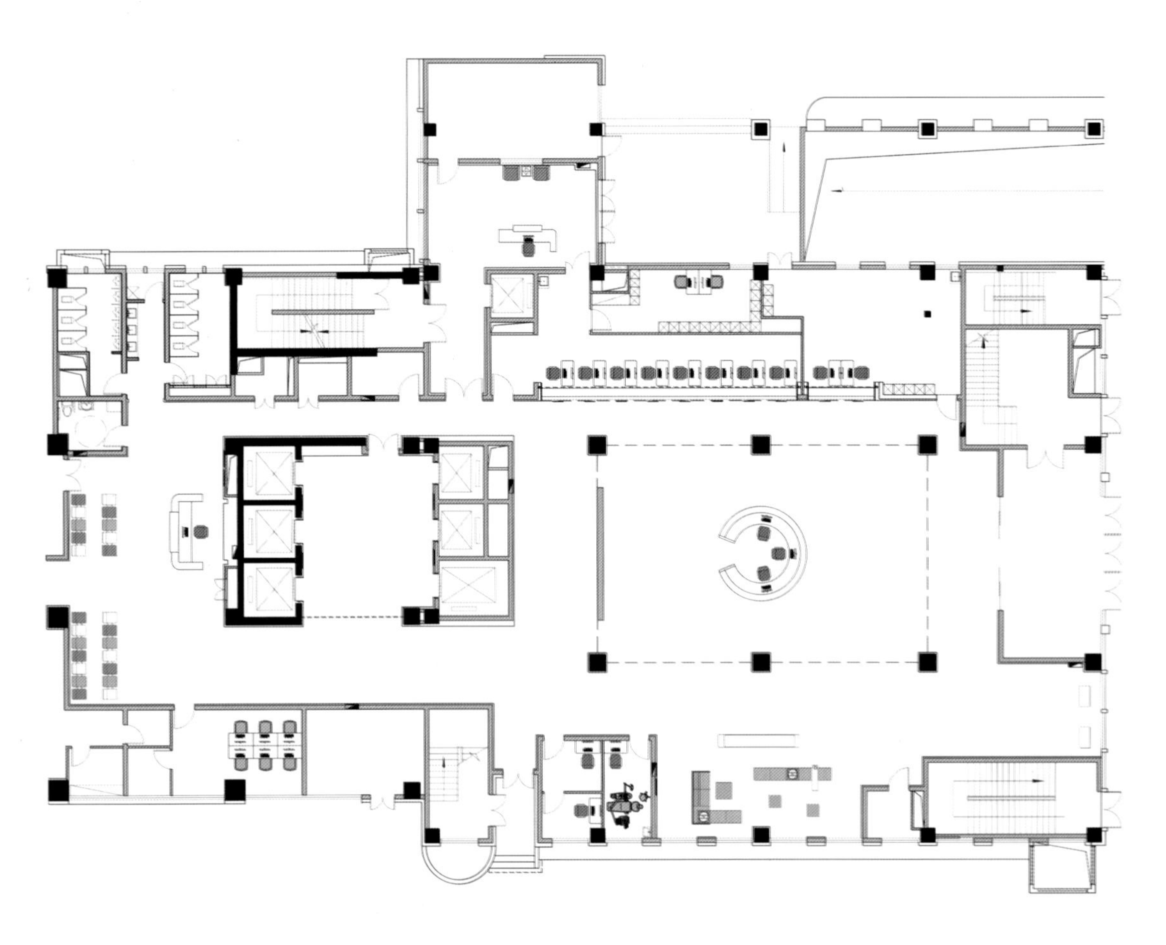

1
Area Index
取药窗口
出入院办理
1974
医学系
口腔专业
口腔专科医院
WE ARE
HOSPITAL
CAREFULLY
精益求精
口腔专科医院
RESPECT PEOPLE DEDICATED
TO FIND SINCERITY PEOPLE 我们

01
08 挂号
Registration
07 挂号
1
Area index
取药窗口
出入院办理
Registration
HE LIGHTS UP
A NEW ERA
WE ARE
HOSPITAL

南京医科大学附属口腔医院，建筑面积 44000m^2。以白、灰、绿、蓝为色调的设计风格，呼应自然健康的主题，候诊大厅内的白色支柱打造出棱线感，寓意健康的牙齿。给患者增加康复的信心。

Jiangsu Province Stomatological Hospital, the affiliated hospital of Nanjing Medical College has a building area of 44,000 square meters. The tones of white gray, green and blue correspond with the theme of nature and health. The pillars in the waiting hall create a ridge feeling suggesting good teeth, which increases the patients' confidence in recovery.

01
取药
Take medicine
Registration
1

3F
4
3
2
1

VIP

Nurse Station

护士站
nurse station

Clinic
01
Clinic
02
Clinic
03

04
贵宾诊室
VIP Clinic
06
贵宾诊室
VIP Clinic

江苏省南京医科大学附属逸夫医院

Sir Run Run Hospital Affiliated to Nanjing Medical University

南京医科大学附属明德医院，是由南京医科大学与江宁区人民政府共同合作建立的三级甲等综合医院，现已更名为南京医科大学附属逸夫医院。去繁除杂是该院的设计原则，以简约的设计语言和界面，深化空间。设计时将水元素以多种形式运用其中，“为有源头活水来”，以看似简单的水带出“生命源头，人文精神”的丰富内涵。

Sir Run Run Hospital Affiliated to Nanjing Medical University, formerly Mingde Hospital Affiliated to Nanjing Medical University, is a joint comprehensive class-A hospital of people's government of Jiangning district and Nanjing Medical University. Water, one of its design elements, which is applied in a variety of forms, carries rich connotations like "the origin of life is the humanistic spirit".

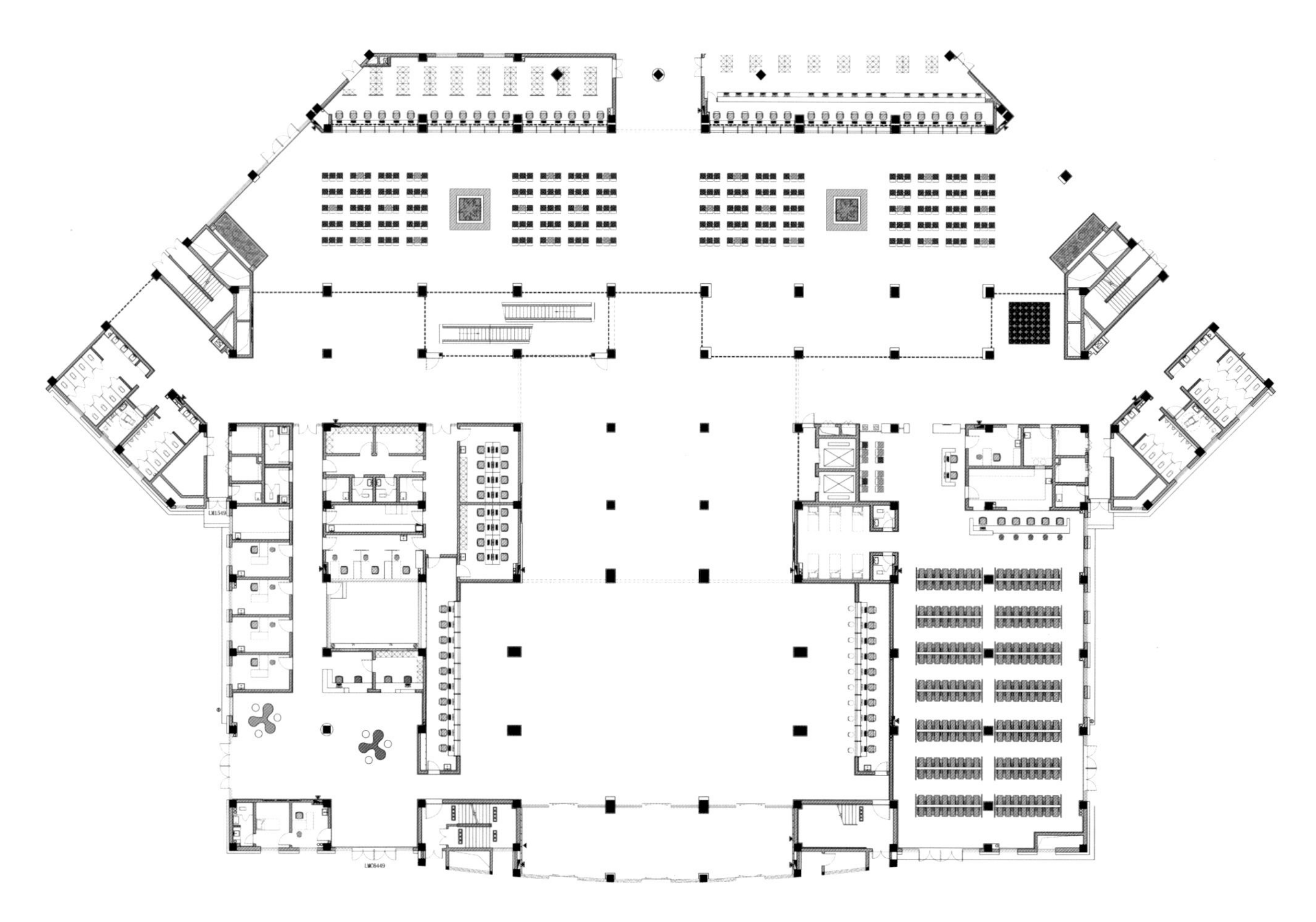

儿科护士站

该院的主医街交汇处设置了水景以及绿植，水声潺潺，绿意柔柔，有助于安抚病患的负面情绪。

Waterscape and green plants are arranged at the crossing of the main streets in the hospital. The ripple of brook and the tenderness of grass are helpful in alleviation of the patients' bad mood.

急诊
EMERGENCY CALL!
NURSE STATION
1F
挂号窗口
Register window
取药窗口
Pharmacy windows

放射影像科 Radiological Images

1F
放射影像
Radiological Images
HEALTH
CARE

南京医科大学附属逸夫医院，建筑面积 92081 m^2，拥有床位 1200 张。设计解决了病房中家属与病患共处一室的不便问题。病房走道内采用间接照明的方式，不仅将光效损耗降至最低，也解决了医患行走时出现眩光造成的安全隐患。

Sir Run Run Hospital Affiliated to Nanjing Medical University with a building area of 92,081 square meters opens 1,200 beds. The design solves the problem of inconvenience caused by the patients and their relatives in the same ward. Indirect lighting is adopted in the corridor, which not only minimizes the luminous efficiency loss, but also avoids the risk of dazzling light while the patients are walking.

702

中国医学科学院南京皮肤病医院

Nanjing Dermatology Hospital of Chinese Academy of Medical Sciences

中国医学科学院南京皮肤病医院 1954 年创建于北京，直属中央卫生部领导。1984 年迁至南京，是我国最早成立的从事皮肤病等防治教学为一体的国家级专业机构。在该院设计中门诊服务台的背景立面、大堂背景立面、门诊室木门等部位均运用了大量的钟山梅花元素，并以活水与游鱼的水景设置，来体现万物之生命力。

Nanjing Dermatology Hospital of Chinese Academy of Medical Sciences, founded in Beijing in 1954, is directly subordinated to the Ministry of Health, and removed to Nanjing in 1984. It is the first state-level specialized educational institute of dermatosis prevention and control. In its design, a great many Zhongshan plum blossoms are employed in the façade of reception desk and lobby and the wooden doors of out-patient rooms, and the waterscape of flowing water and swimming fish exhibits vitality of the nature.

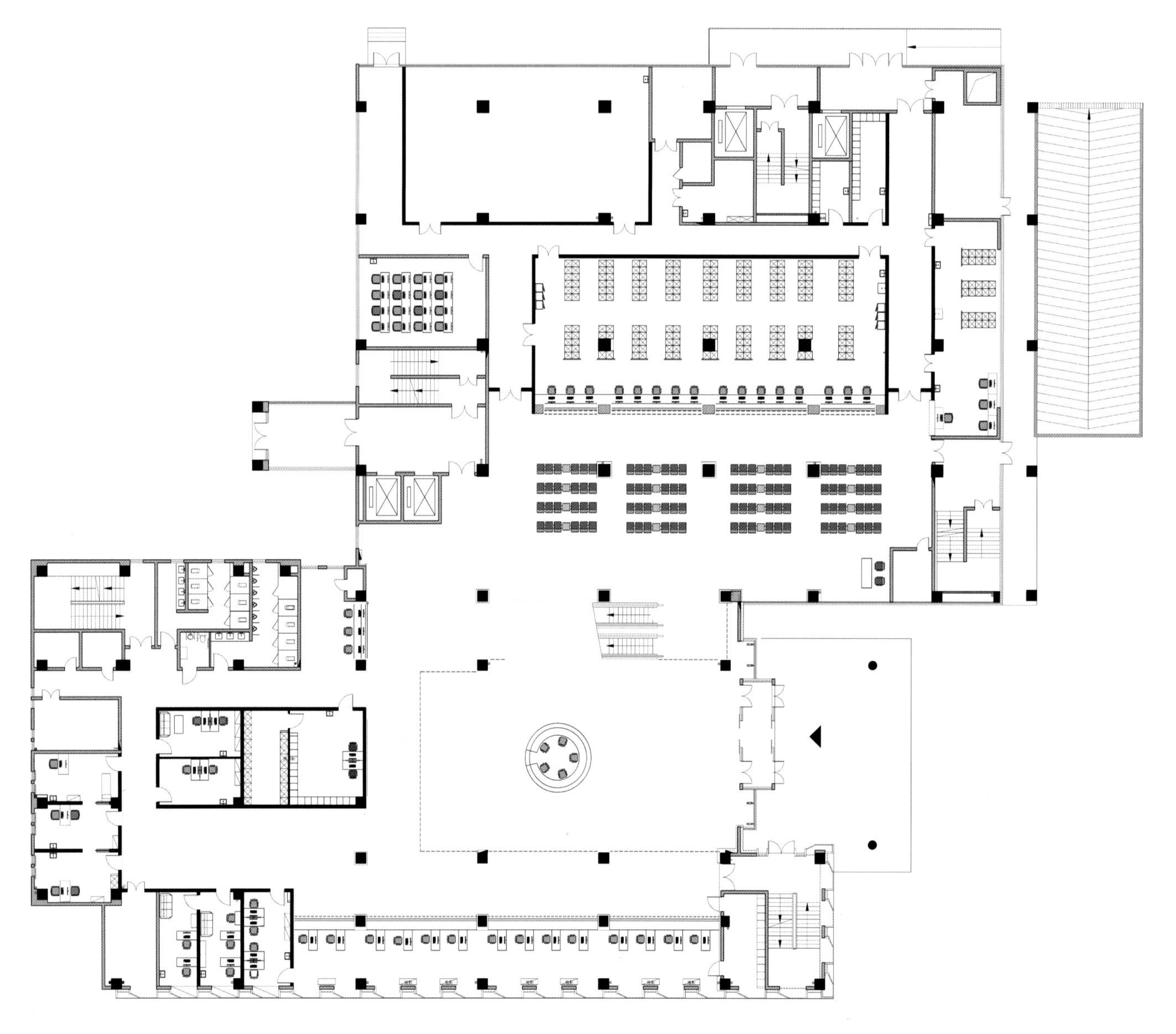

普通门诊
Average outpatient service

02收费
1F
Floor Information
注射区
病历资料室
01收费
02收费
03收费

1F
1F
1F
护士站 / NURSE STATION

中国医学科学院南京皮肤病医院，建筑面积 30611 m^2，两个病区设有编制床位 100 张。空间多使用光滑简洁的材料，主色调为让人产生亲近感的白与绿。采用柔和灯光及泛光灯光，降低皮肤病患者对医院空间的敏感度。

Nanjing Dermatology Hospital of Chinese Academy of Medical Sciences with an area of 30,611 square meters opens two ward areas with a total of 100 beds. Smooth, simple and clean materials are mainly used in the space, and the colors of white and green are the dominant tone giving friendliness and intimacy.

01收费
05收费
04收费
information
1F
Floor Information
注射区
病历资料室

皮肤外科 整形科
Skin surgery ZhengXingKe

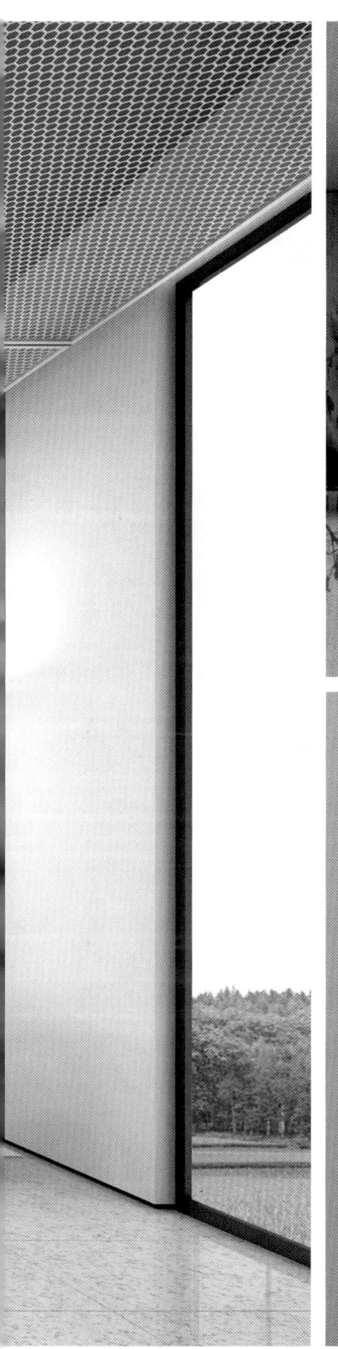

特需候诊区
Waiting area special
特需候诊区 Waiting area special

江苏省南京南部新城医疗中心（方案）

New Urban Medical Center in Southern Nanjing (Proposal)

南京南部新城医疗中心，是正兴建于秦淮区的一家现代化三级甲等大型综合医院。秦淮区是包含夫子庙、古秦淮河在内的南京历史底蕴最为丰厚的区域之一，自然而然，在这座医院的设计中，中医文化、中国书法、老城南的特有城市元素及内涵成为空间设计的主要载体。医院空间氛围融入南京特有的城市人文情怀，塑造出独树一帜的本土化空间。

New Urban Medical Center in Southern Nanjing is a modern, comprehensive class-A hospital with a large building area under construction in Qinhuai district, an area including the Confucius Temple, ancient Qinhuai River, etc. with the most generous history. Naturally, the traditional Chinese medicine, Chinese calligraphy and unique traits of southern Nanjing and its connotations become the carrier of its space design. The hospital's atmosphere is immersed in the city's humanistic feelings, thus a distinctive local space is formed.

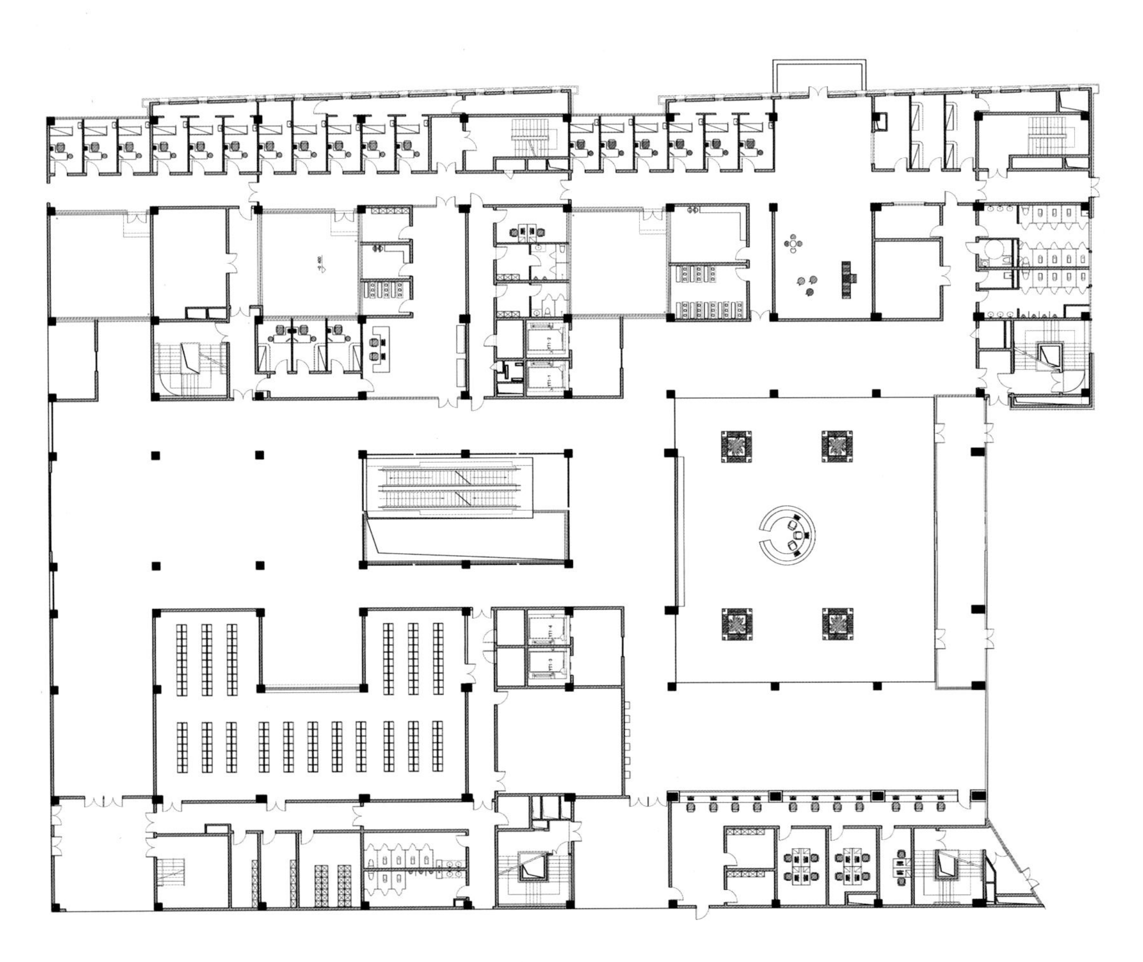

本草纲目

千年中醫

取药
Take medicine
03 取药
Take medicine
04 取药
Take medicine

南京南部新城医疗中心，建筑面积 307000m²，设置床位 1500 张，服务台设置了放包台，便于减轻病患负担，无障碍化卫生间设计，让行动不便的病患更有尊严的使用这一空间。

New Urban Medical Center in Southern Nanjing with a building area of 307,000 square meters opens 1,500 beds. The reception desk is specially designed for handbags or parcels so as to ease the burden of patients. The design of accessible toilet for the handicapped also offers more dignity to the patients.

1F
一次候诊
Waiting Area
①城市医保
②新农合医保
③自费医疗
④缴费
升降梯
卫生间
自动扶梯
消防楼梯
残卫

创新设计的围合输液椅，可以给病患以及陪护提供一个独立安静的区域。设置的抽屉可放置输液药水以及个人物品，更具人性化考量。
使用半墙式隔屏取代传统的隔帘，在保护病患隐私的同时，传统人文气息的隔屏与空间相得益彰。

The innovation of infusion chair offers a quiet and private area to the patients and their company. The drawer, with humanistic concern, is arranged to place medicine and personal stuff.
The usual curtains are replaced by half-wall partitions, which is aimed to protect patients' privacy whilst these traditional style partitions achieve complementarities with the overall space.

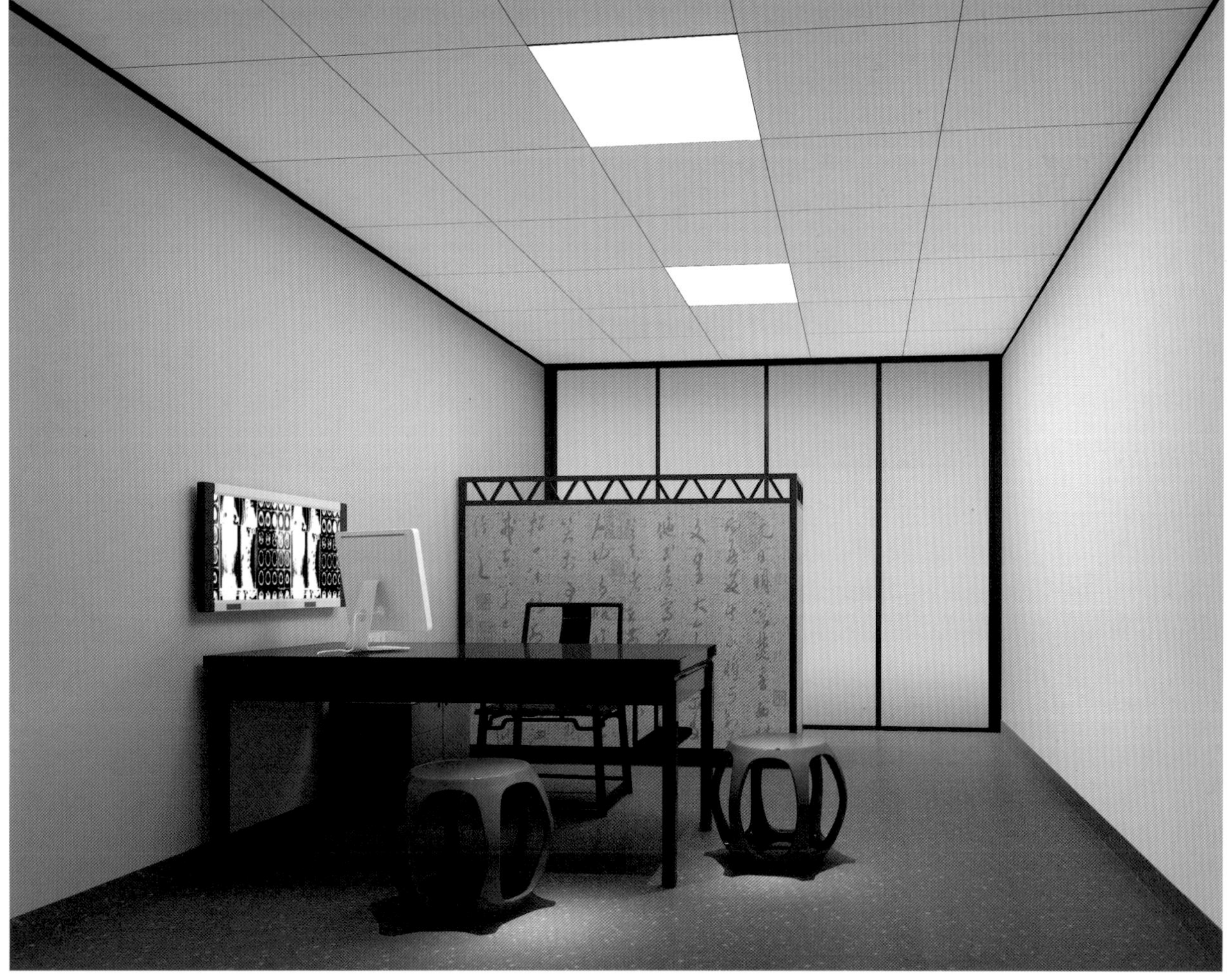

名醫堂

1F
出入院大厅
Entrance hall

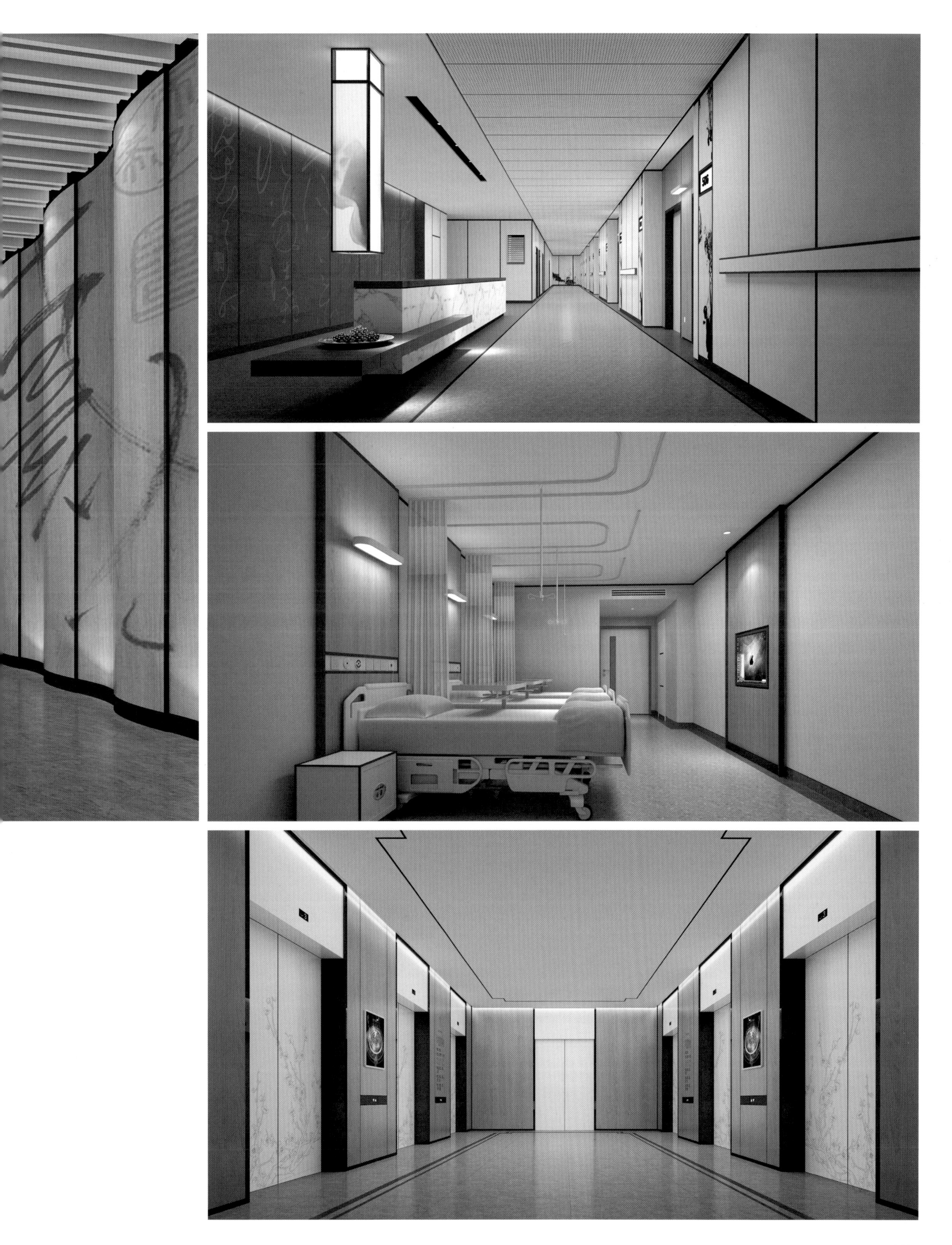

南京医科大学第二附属医院

The Second Affiliated Hospital of Nanjing Medical University

东侧倚靠秦淮小桃园湖畔的南京医科大学第二附属医院是三级甲等综合医院，设计中选用了翠竹、山石、流水等自然元素，让医院空间更具亲和力。为了贴合医院现代化、数字化、国际化医疗中心的定位，设计中将分子结构、DNA等元素形象化后运用于空间，以体现生命科学的概念。

The Second Affiliated Hospital of Nanjing Medical University is a comprehensive class-A hospital leaning against the Qinhuai River on its east side. Natural elements like bamboo, rock and flowing water are employed in the design, which bring more affinity to the hospital. In order to fit its modern, digital, and international medical orientation, elements such as molecules and DNA are visually presented, interpreting the definition of biology.

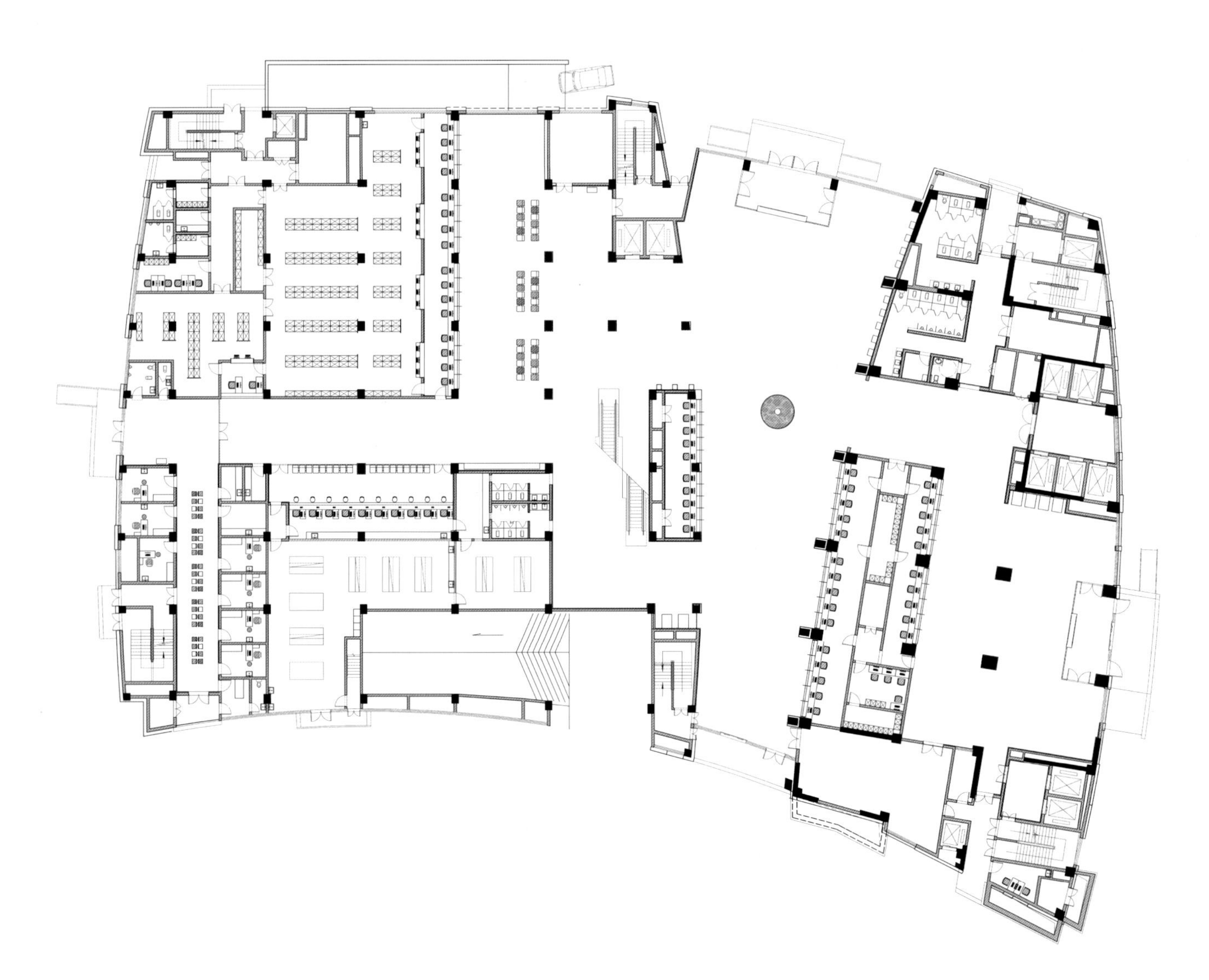

厚德博学
生命至上

1
Area Index

端景鱼缸增加了医院空间的互动性，取药窗口采用隔断隔开，并设放包台，保护每一位病患的隐私。

The embedded fish tank, as the symbol of good fortune, adds more interactivity in the hospital space. The service counter for medicine taking is separated by partitions, and is specially designed for handbags or parcels, which protects the privacy of each patient.

13
取药
Take medicine
12
取药
Take medicine
取药
03
02
01
Take medicine

南京医科大学第二附属医院，建筑面积 76400m^2，设有床位 1200 张。设计充分解决了医院顶面杂乱的问题，仅以一种材料打造鳞次栉比的空间感，并将灯具完美嵌入顶面。
灯具采用柔和及泛光照明，避免医护人员长时间在灯光下工作带来的视力伤害。

The Second Affiliated Hospital of Nanjing Medical University with a building area of 76,400 square meters opens 1,200 beds. The design makes the ceiling clear of clutter by the orderly arrangement of one single material, with lighting perfectly embedded.
Floodlighting is adopted and the soft light avoids harm to the medical staff in their long-time work.

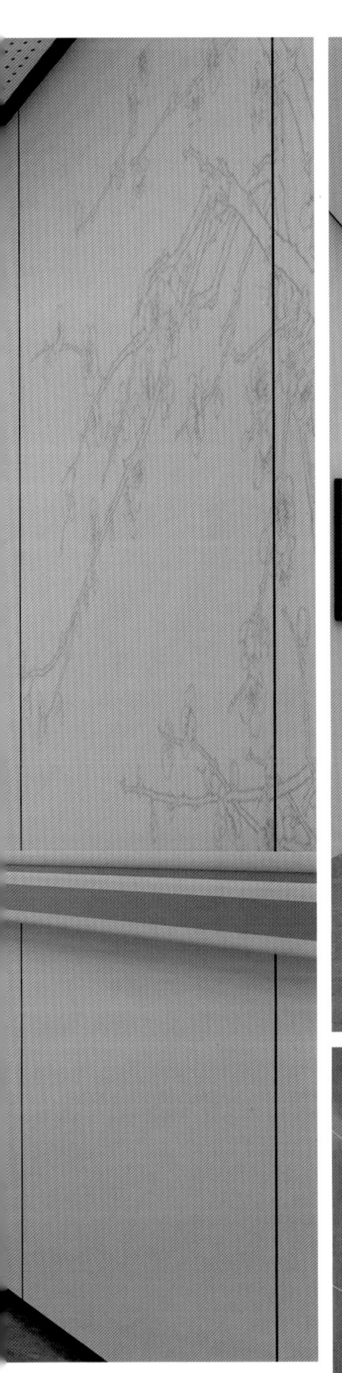

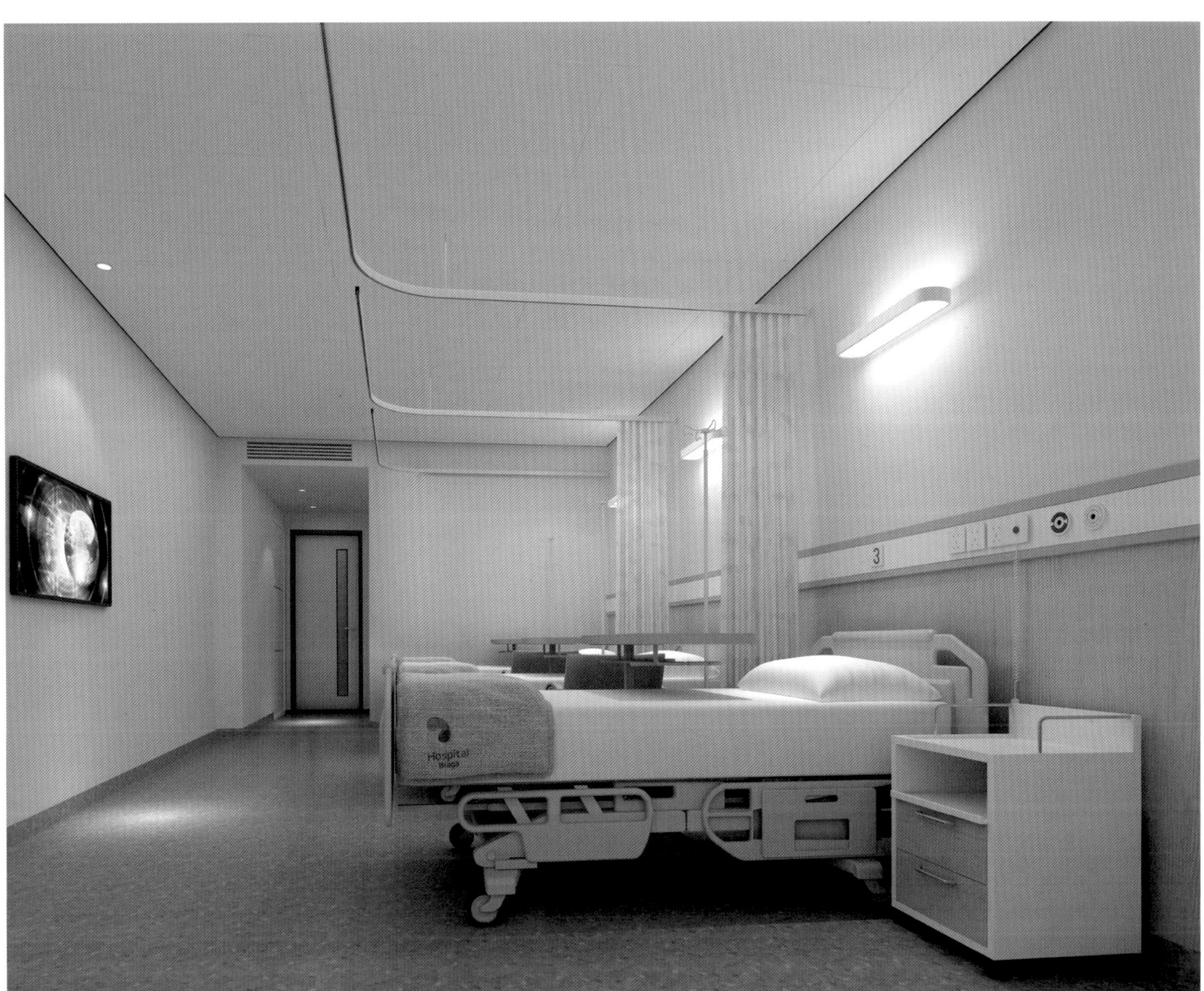
3
Hospital

江苏省泰州市人民医院

Taizhou People's Hospital, Jiangsu

南唐时，泰州为州治，取“国泰民安”之意，始名泰州。泰州市人民医院始建于 1917 年，最初为美国基督教教会创建的福音医院，2011 年经原江苏省卫生厅批准成为国家三级甲等综合性医院。本次设计以“脉”为概念，“左月右永，月者肉也，永者水也。”与医疗空间完美契合。其市树银杏有长寿之星之说，且叶状似心形。在本设计中被用于叶脉这一设计元素，另有经脉、文脉共三种，彰显其人文特色。

Taizhou, as the implication of its name–peace and prosperity, was a state when under the reign of Southern Tang. As early as in 1917, Taizhou People's Hospital was originally founded as a Christian hospital sponsored by the U.S. Christian Church. As a comprehensive class-A hospital, Taizhou People's Hospital was approved by the former Health Department of Jiangsu province in 2011. The design of this case interprets the Chinese character " 脉 ". This character derives from " 月 ", meaning "flesh" and " 永 ", meaning "forever". In the language of Chinese, " 脉 " refers to arteries and veins of humans or the vein of leaves or insects. Therefore, the vein of heart-shaped leaves of maidenhair trees, a symbol of longevity in Chinese tradition and also known as the city tree, is one of the three design elements in the hospital, and the other two are meridian and unity or coherence in writing, a manifestation of its humanistic features.

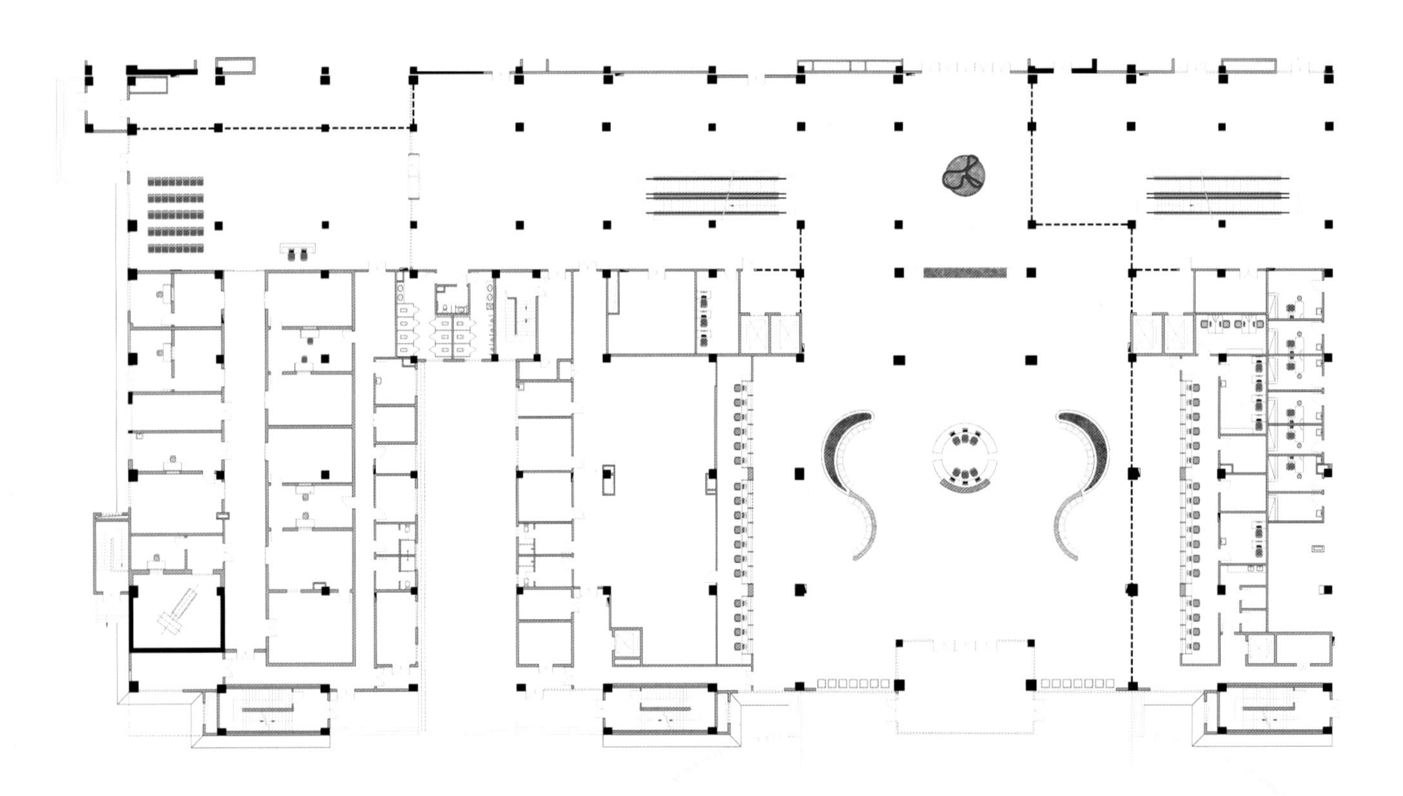

1F
1F

厚德博学
生命至上
取药
取药
1F

03 出入院办理
04 出入院办理
05 出入院办理
06 出入院办理
厚德博
03
04
05
06

顶部照明融入叶脉设计元素，通道相交设置鱼缸，等候椅与绿植相结合，意在营造令人放松的就医氛围。

To create a relaxing and pleasant hospital environment, the design of the vein of leaves is adopted in top lighting; fish tanks are arranged at passage crossings and green plants are coupled with waiting chairs.

中庭的绿植与休憩的搁板合二为一拉近了人与自然之间的距离。灯具采用医学分子造型，衬托生命科学的设计概念。
输液区采用围合座椅，给予病患一个相对隐私的空间，避免受到其他病患的打扰。

Green plants and shelves are combined in the central area of the hospital, narrowing the distance between nature and humans. Lamps are shaped in the form of molecules, serving the design concept of biology.
The innovation of infusion chair offers a comparatively quiet area to the patients in case of interruptions nearby.

挂号收费
挂号收费
01
02
Place
HOSPITAL

护士站
B1
A1
A2

护士站 Nurse Station

05
Cardiology
Department

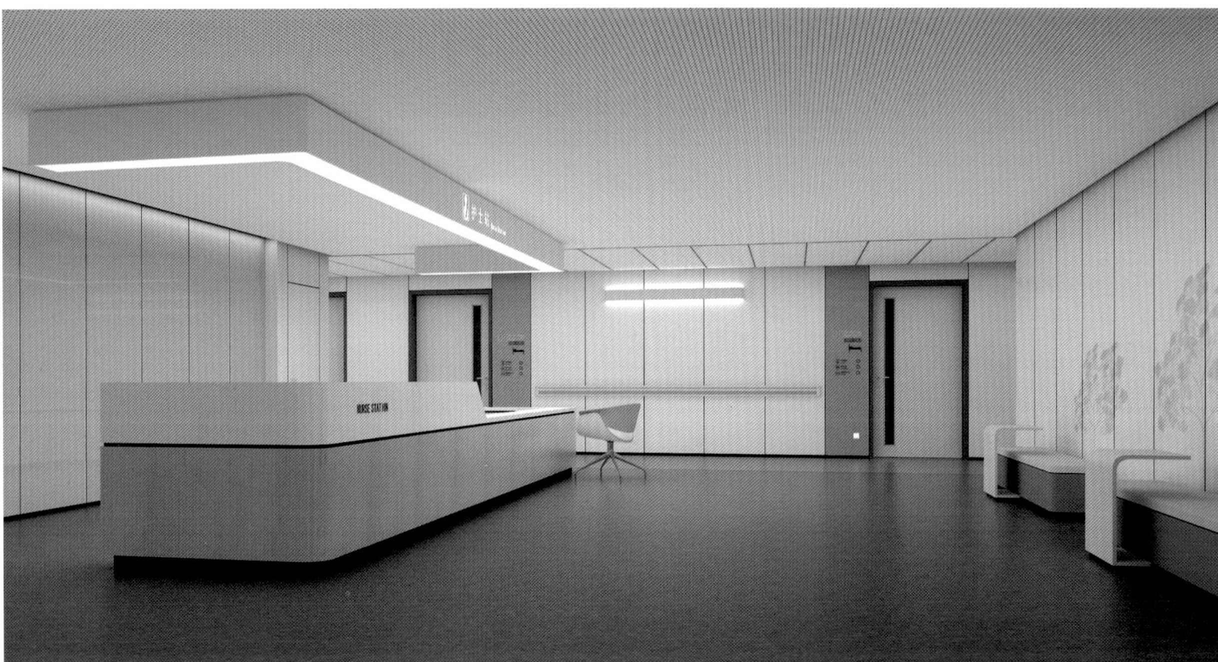
NURSE STATION

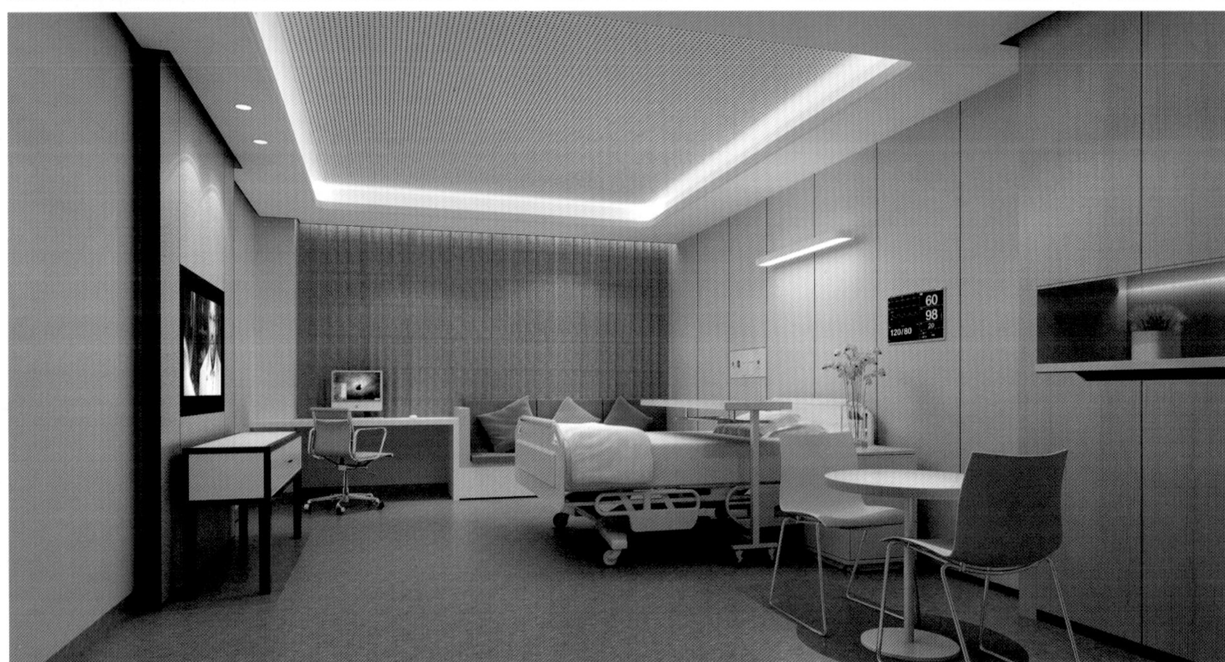
60
98
120/80

3
3
1

泰州市人民医院建筑面积 295000m^2，设有床位 3600 张。作为一家规模可观的大型医院，泰州市人民医院的设计做到了流程合理化、功能人性化、环境舒适化，并将百年人文底蕴与人文关怀得以继承绵延。

Taizhou People's Hospital with a building area of 295,000 square meters opens 3,600 beds. As a comprehensive hospital of considerable scale, its design achieves propriety in the procedure, humanity in the function and comfort in the environment, and therefore, humanistic connotation and care can be inherited and passed on.

医学开

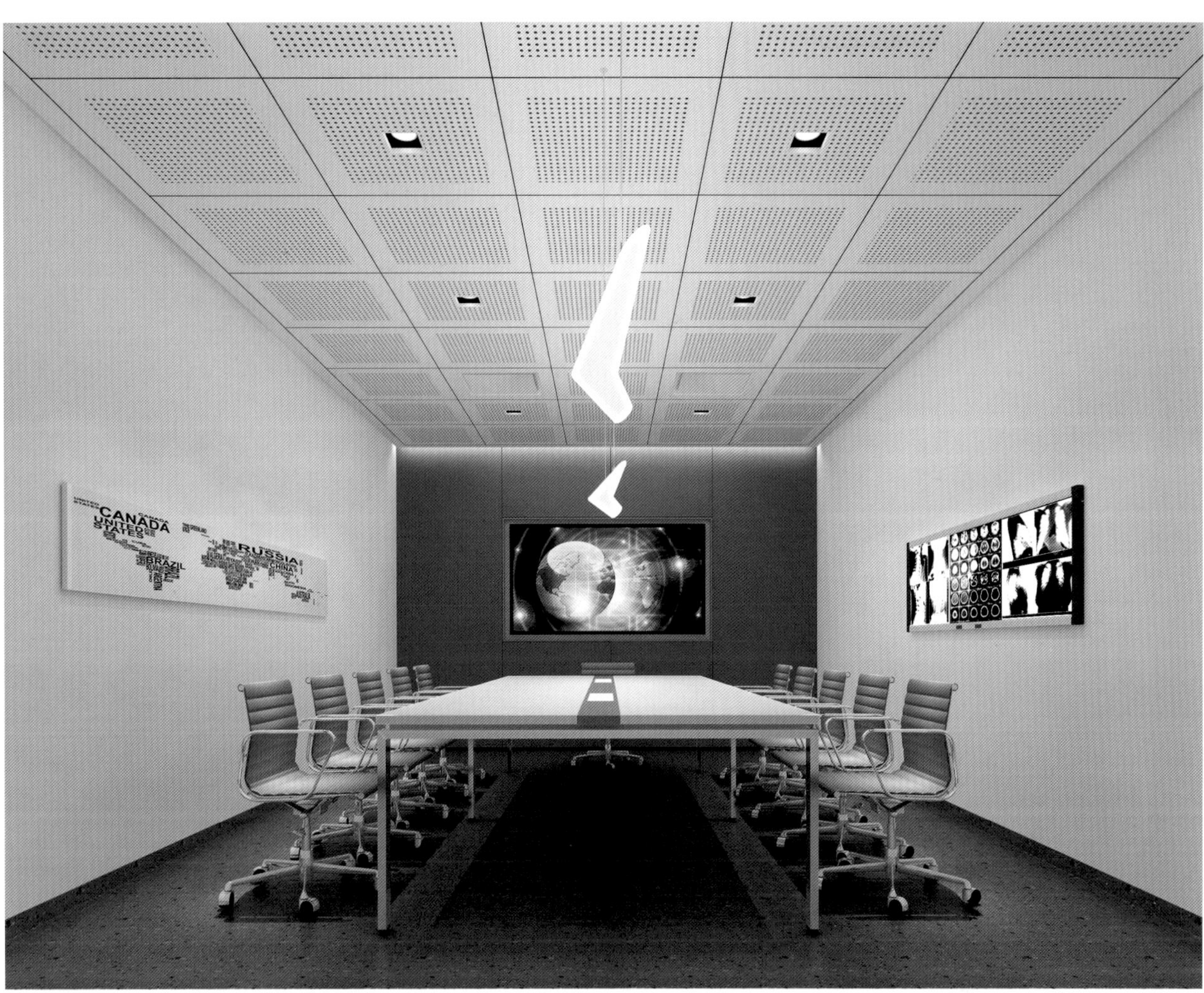
CANADA
UNITED STATES
RUSSIA
BRAZIL

淮安经济技术开发区人民医院

Huai'an Economic & Technological Development Zone People's Hospital

淮安经济技术开发区人民医院，是为满足该地区就医问题所新建的一家综合性医院。本次设计以“星空下·山水间”为概念与医疗空间完美契合。以湖为生的游艇以及浩瀚的星空元素，在设计中将其进行艺术提炼引入医院大堂中。

Huai'an Economic & Technological Development Zone People's Hospital is a newly - founded comprehensive hospital aiming to fulfill the local need for medical service. Following the idea of "a hospital under the stars and amid the landscape", the design is perfectly matched with the hospital space.

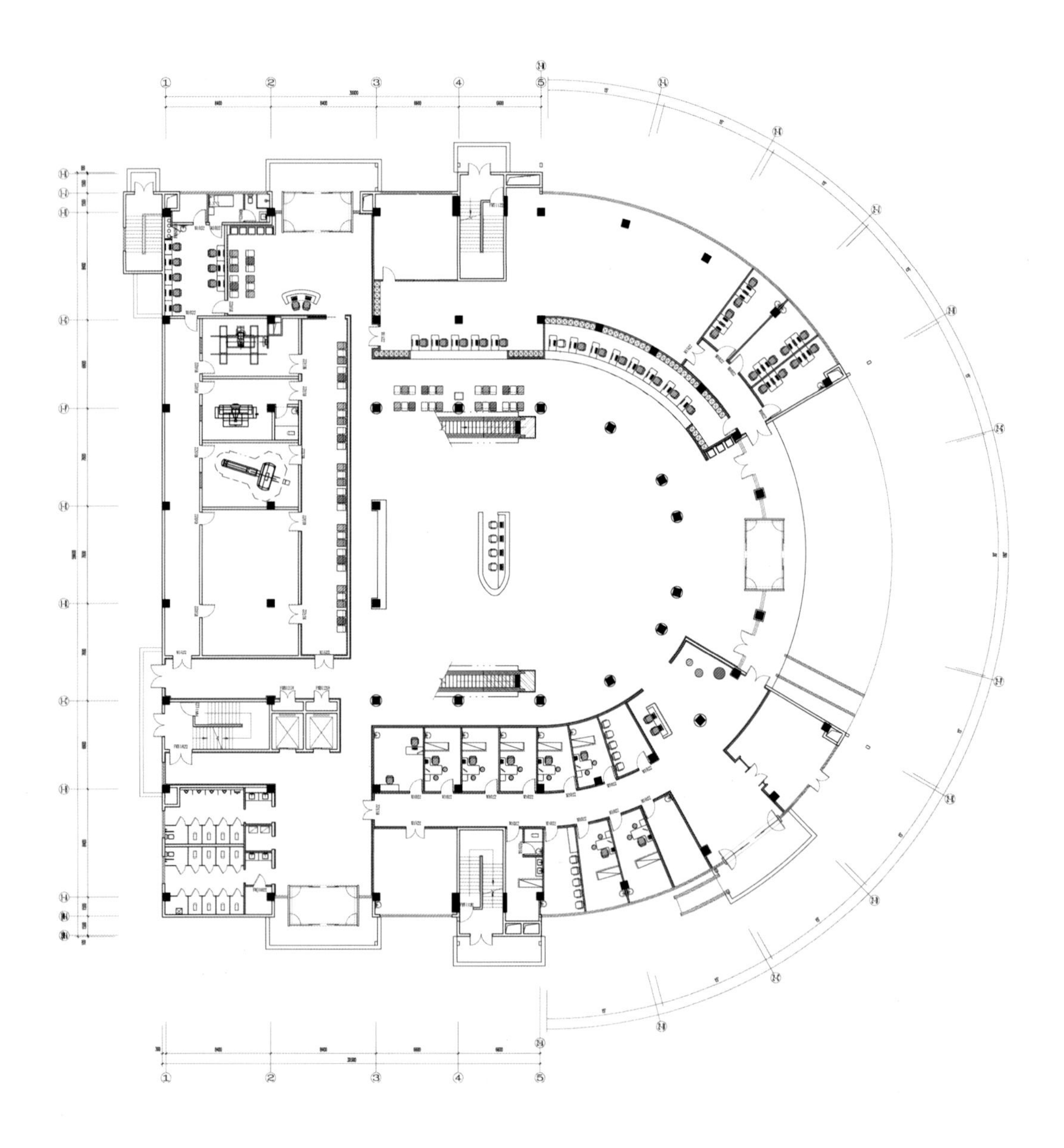

护士站
urse Station

Trigger point block

开发区
院
technological
hospital
Information
1F

护士站高低台的设置更能体现人性化。等候区用木地板划分，让患者感受到家的温暖。

Humanity finds good expression in double-level counters, and the waiting area is isolated by wooden floor, which makes the patients feel ease at home.

诊室6

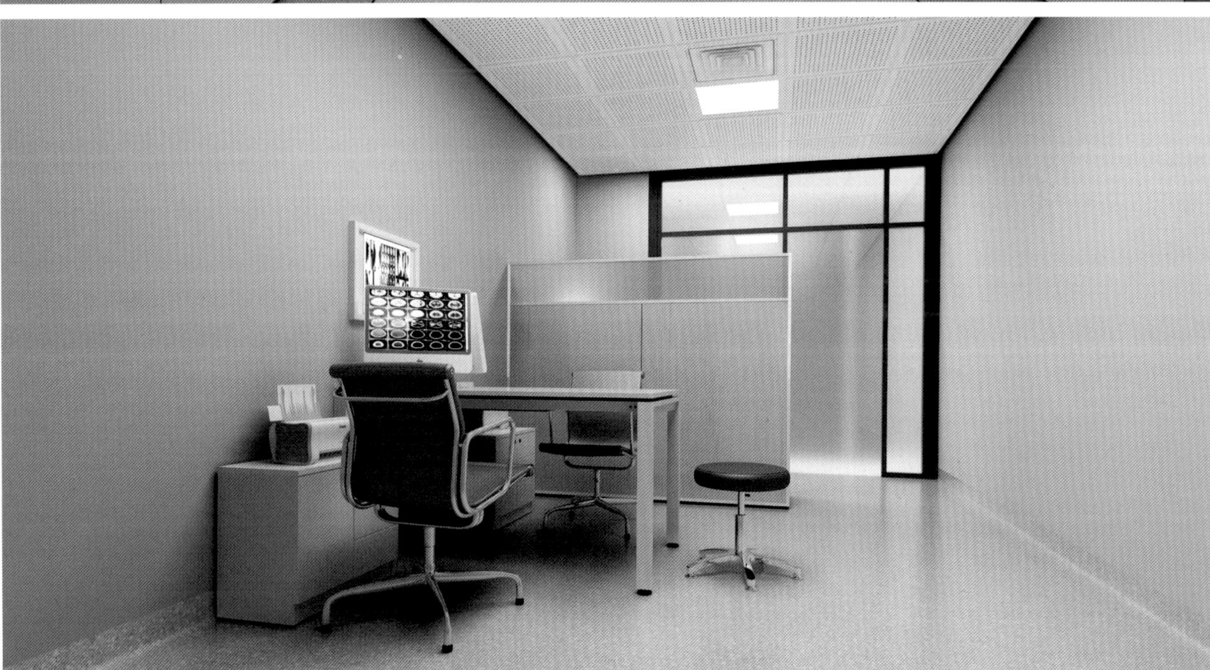

NURSING STAITON

淮安经济技术开发区人民医院建筑面积 17990m²。作为一家新兴的综合性医院，淮安经济技术开发区人民医院的设计做到了流程合理化、功能人性化、环境舒适化。

水景让患者感受与自然的亲近，母性关爱的雕塑使空间更加温馨。

The building area of Huai'an Economic & Technological Development Zone People's Hospital is 17,990 square meters. As a newly developed comprehensive hospital, it achieves propriety in the procedure, humanity in the function and comfort in the environment.

The waterscape inspires the patients to get close to nature, and the sculpture with the implication of maternal love makes the space warm and sweet.

护士站 /Nurse Station
906
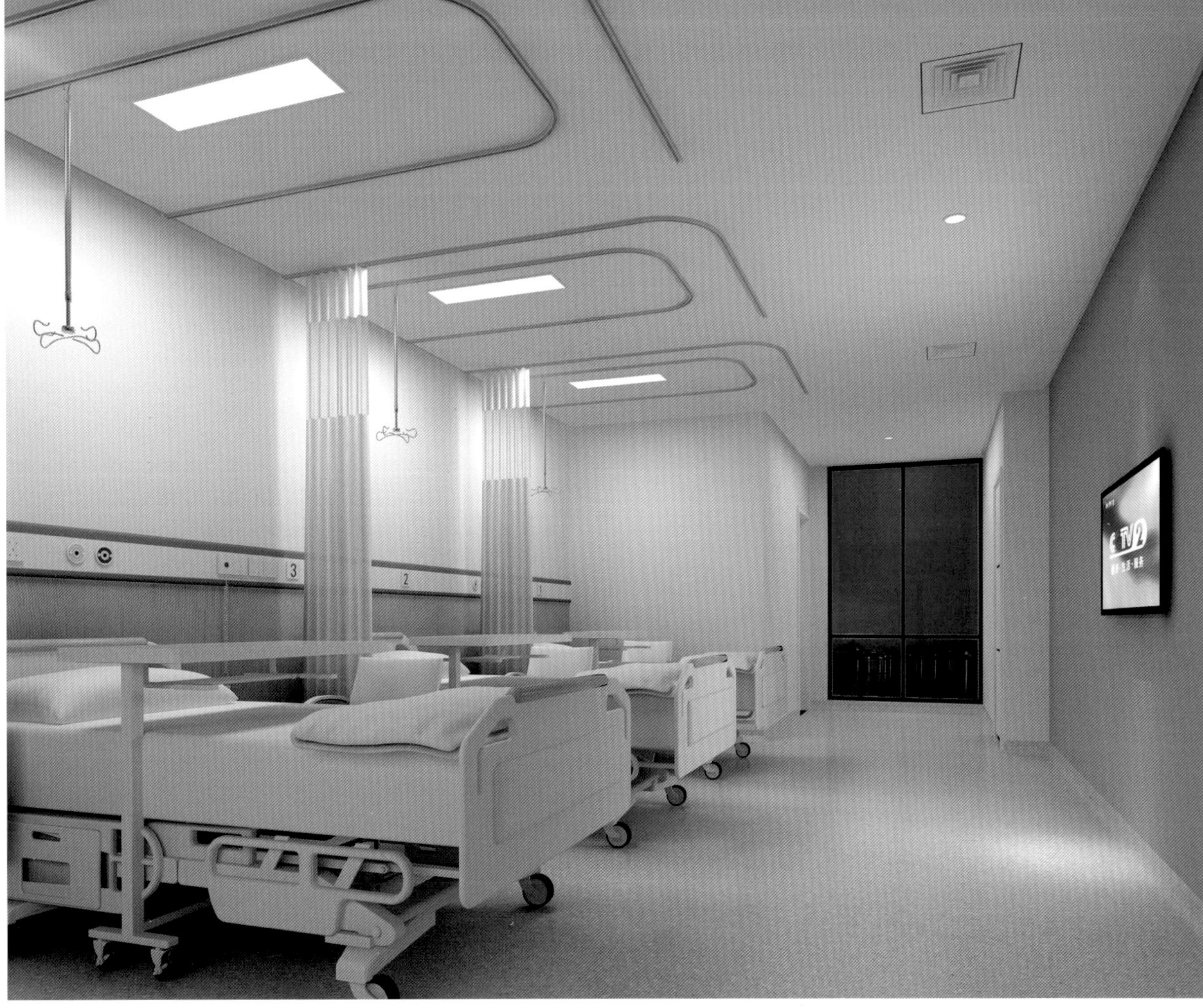

医学开拓健康研讨会
2012 Medicine to develop health conference

南通大学附属医院

Affiliated Hospital of Nantong University

南通大学附属医院由清末状元张謇先生创建于 1911 年。1994 年被卫生部首批评定为三级甲等综合性医院，1998 年被确定为国际紧急救援中心网络医院。本次设计以“星空下・山水间”以及医院门口的“古银杏”为概念与医疗空间完美契合。在设计中将其进行艺术提炼引入医院大堂中。

Affiliated Hospital of Nantong University was founded in 1911 by Zhang Yu, an imperial examination champion in late Qing dynasty. In 1994, it was graded as one of the first group of class-A hospitals by the Ministry of Health. In 1998, it is identified with international emergency aid center online hospital. Following the idea of "a hospital under the stars and amid the landscape with old ginkgo trees at its entrance", the design is perfectly matched with the hospital space.

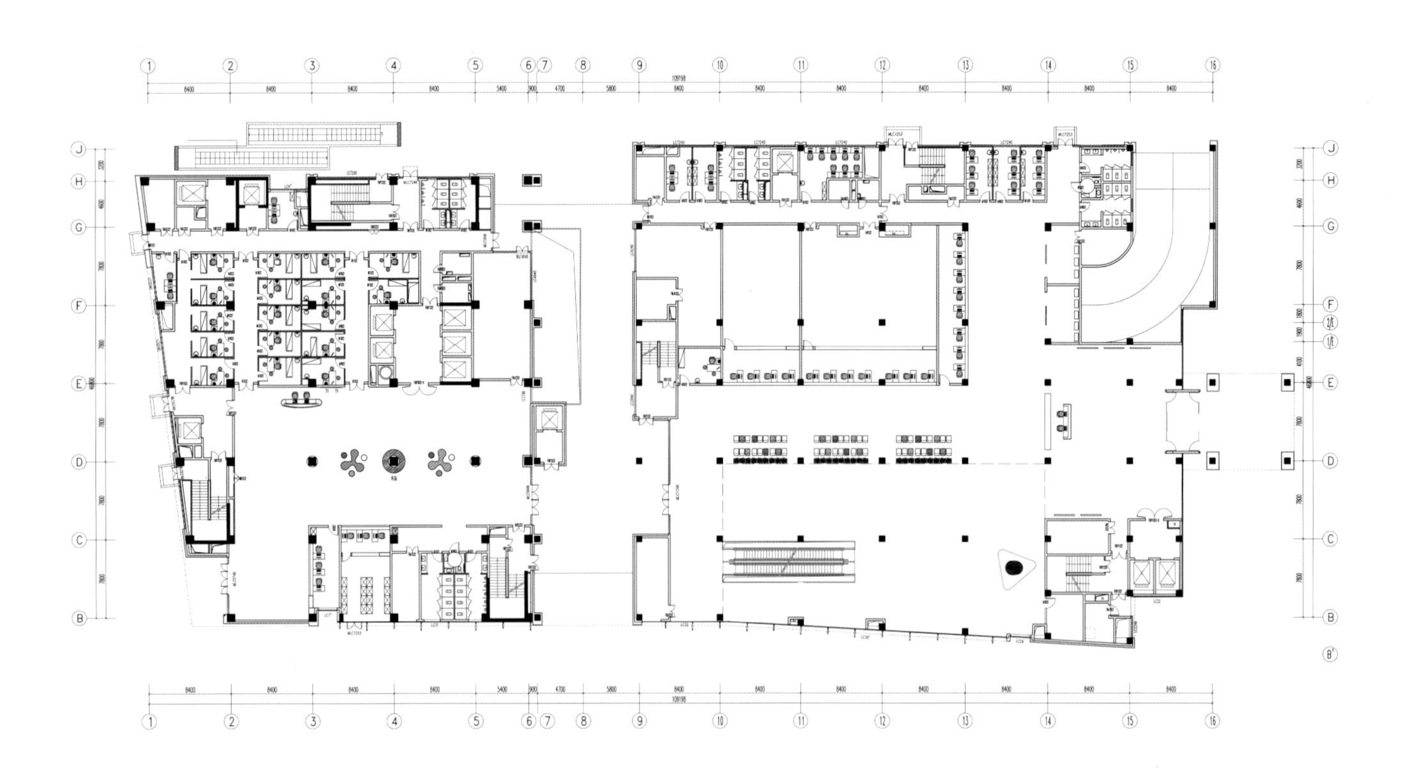

大医精
INFORMATION DESK

慈善
02

围绕圆柱设计的鱼缸，给予孩子和大自然的互动。

Embracing the pillar, the fish tank offers children interactivity with the nature.

01 取药窗口
02 取药窗口
03 取药窗口
04 取药窗口

与庭院结合的侯诊区，让患者舒适等候的同时，更能感受自然的美。

Associated with courtyard, the waiting area makes the patients feel comfortable and the beauty of nature as well.

05 挂号收费
04 挂号收费
03
04

南通大学附属医院门诊楼建筑面积40188m²。作为江苏省“科教强卫工程”十大“临床医学中心”之一，南通大学附属医院的设计做到了流程合理化、功能人性化、环境舒适化。

The outpatient building of Affiliated Hospital of Nantong University has 40,188 square meters. As one of "Top 10 Clinical Medical Centers" nominated by Jiangsu province in "Project of Invigorating Health Service through Science and Education", the design of Affiliated Hospital of Nantong University achieves propriety in the procedure, humanity in the function and comfort in the environment.

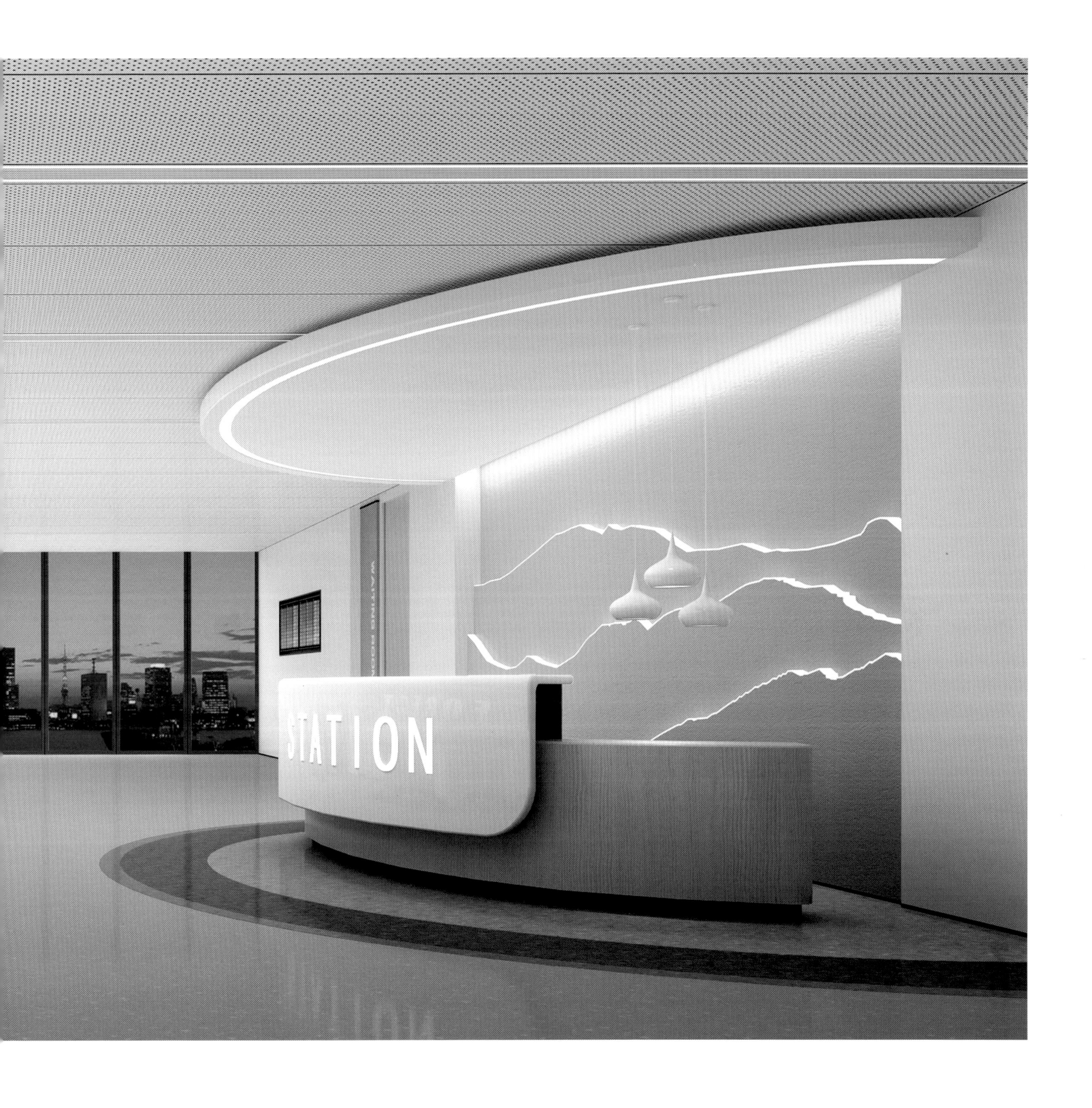
STATION

吉林省四平中心人民医院
Jilin Province Siping People's Central Hospital

有“东方马德里”之名的四平是吉林省的“南大门”，拥有3000年的历史。四平中心人民医院是四平地区唯一一所集医疗、科研、教学、预防保健和康复功能为一体的综合性三级甲等医院。设计中以纯粹凝练的自然元素融入极简空间，形成自然与科技感相容的氛围。将水泡的自然状态进行艺术提炼，具象成照明、灯具，整体色调明朗轻快，对病患及家属的心理起到积极作用。

Siping, the so - called "Oriental Madrid" and known as the "South Gate" of Jilin province, boasts a long history of 3,000 years. Siping People's Central Hospital is the only comprehensive class-A hospital with the integration of medical care, scientific research, education, prevention and healthcare and rehabilitation. The purely condensed natural elements are immersed in the simplified interior space, forming a natural atmosphere compatible with science and technology feeling. The natural state of bubbles is artistically refined and is in concrete forms like lighting lamps. The overall tone of the hospital is bright, clear and lively, making a positive impact on the patients and their relatives.

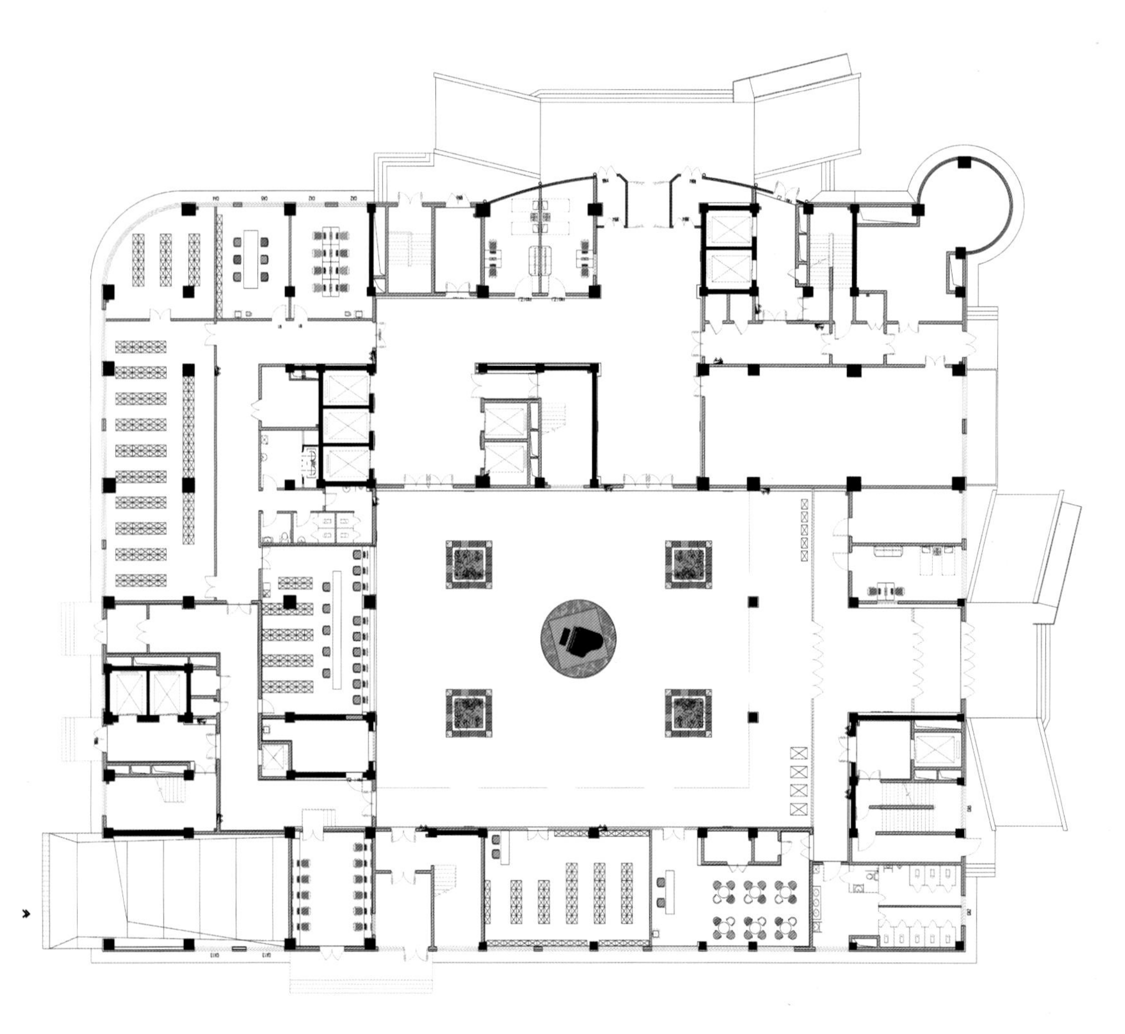

超市

为保护病人隐私，取药窗口处以隔板相区分，等候大厅中，设置与树池相结合的等候座椅，拉近人与自然的距离。

In order to protect patients' privacy, the service windows of Pharmacy are separated by partitions. The waiting chairs arranged in association with cubes for plants, narrowing the distance between human and nature.

09
取药
Take medicine
08
取药
Take medicine
07
取药
Take medicine
03
02
01
Take medicine

四平中心人民医院，建筑面积 57000m²，设有床位 1000 张。在儿外科的室内设计中，动漫元素让就诊患儿降低畏惧情绪，在家具座椅上亦采用悟空的动漫元素，在不影响风格的同时充分做到因地制宜。

Siping People's Central Hospital with a building area of 57,000 square meters opens 1,000 beds. In Pediatric Surgery design, animation elements allay children patients' fear. Moreover, the Monkey King image is applied on the chairs, a successful adaptation to the overall style.

SIPING CENTRAL HOSPITAL
1978

Windows

PEDIATRIC SURGICAL
Nurse Station 儿外科护士站

山东省济南市第三人民医院

The Third Hospital of Jinan

济南市第三人民医院，是一所集医疗、急救、教学、科研、预防、保健于一体的市级综合性二级甲等医院。作为济南市七大区域型医疗中心之一，济南市第三人民医院以清雅与文化兼具的设计特色在当地独具一格。“四面荷花三面柳，一城山色半城湖”，荷花作为济南的标识，在设计中作为辅助元素运用在护士站背景以及电梯、支柱上，整体颇具清新脱俗感。

The Third Hospital of Jinan is a comprehensive Class-B city-level hospital with integration of medical care, emergency treatment, education, scientific research, prevention and health care. As one of the seven regional medical service centers, The Third Hospital of Jinan is distinctive for its design features of elegance and culture. As is portrayed in the poem, Daming Lake is embraced by lotus flowers, and the reflections of surrounding mountains are mostly found in the lake. Therefore, lotus flower, the symbol of Jinan, is applied in the design as a complementary element to the background of Nurse Station, elevators and pillars, which makes a pure and fresh feeling free from vulgarity.

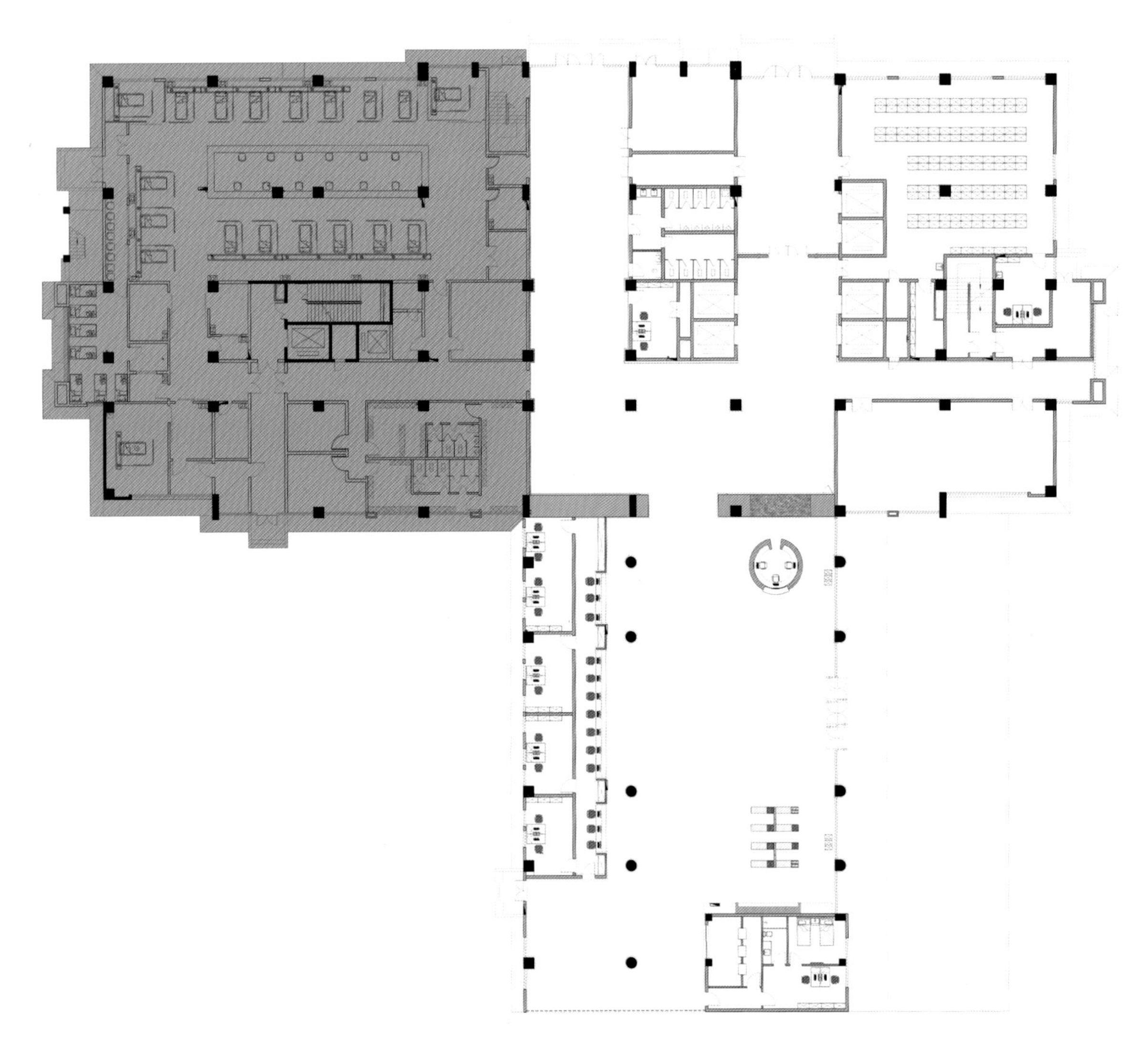

01收费
3收费
04收费
05收费
医院

为了避免空间顶部的灯光、空调风口给人以杂乱感，顶面采用蜂窝板将其隐于其中，在视觉上给人以干净整洁的体验。

Ceiling lights and air-conditioner outlets, the contributors to clutter, are covered by honeycomb panels so that tidy and clean experience is visually provided.

06收费

RECEPTION DESK

济南市第三人民医院，建筑面积 35000m^2，设有床位 1079 张。灰白色占空间色调的九成，绿色以点缀，让原有的医院空间更有人情味与文化内涵。设计上将患者进行流线功能分流，过道电视内嵌入墙，避免人流来往之间造成意外。

The Third Hospital of Jinan with a building area of 35,000 square meters opens 1,079 beds. The main tone of grayish white color covers 90 percent of the space; the green ornaments add more human touch and cultural connotation to the hospital. The TV sets are embedded in the wall, which lower the risk of accidental crushes because patients are linearly grouped in the hospital.

护士站
Nurse Station

湖南省襄阳市襄州区人民医院

Xiangyang Xiangzhou District People's Hospital

襄阳市襄州区人民医院，是 所集医疗、救护、保健为一体的二级甲等综合性医院。位于历史悠久的楚汉文化发源地，襄阳市襄州区人民医院的设计在自然人文的风格上追求简练整洁。运用了襄阳的市树女贞树、以湖为生的游艇以及医院特有的生物分子结构元素，其中女贞树有象征生生不息的生命寓意，在设计中将其进行艺术提炼引入医院大堂中。

Xiangyang Xiangzhou District People's Hospital is a comprehensive class-B city-level hospital with the integration of medical care, emergency aid and health care. Located in the cradle of Chu-han culture, the design of Xiangzhou District People's Hospital pursues simplicity and tidiness. Therefore, Chinese glossy privets, the symbol of ever - growing vitality, yachts and medical molecules are employed in its design.

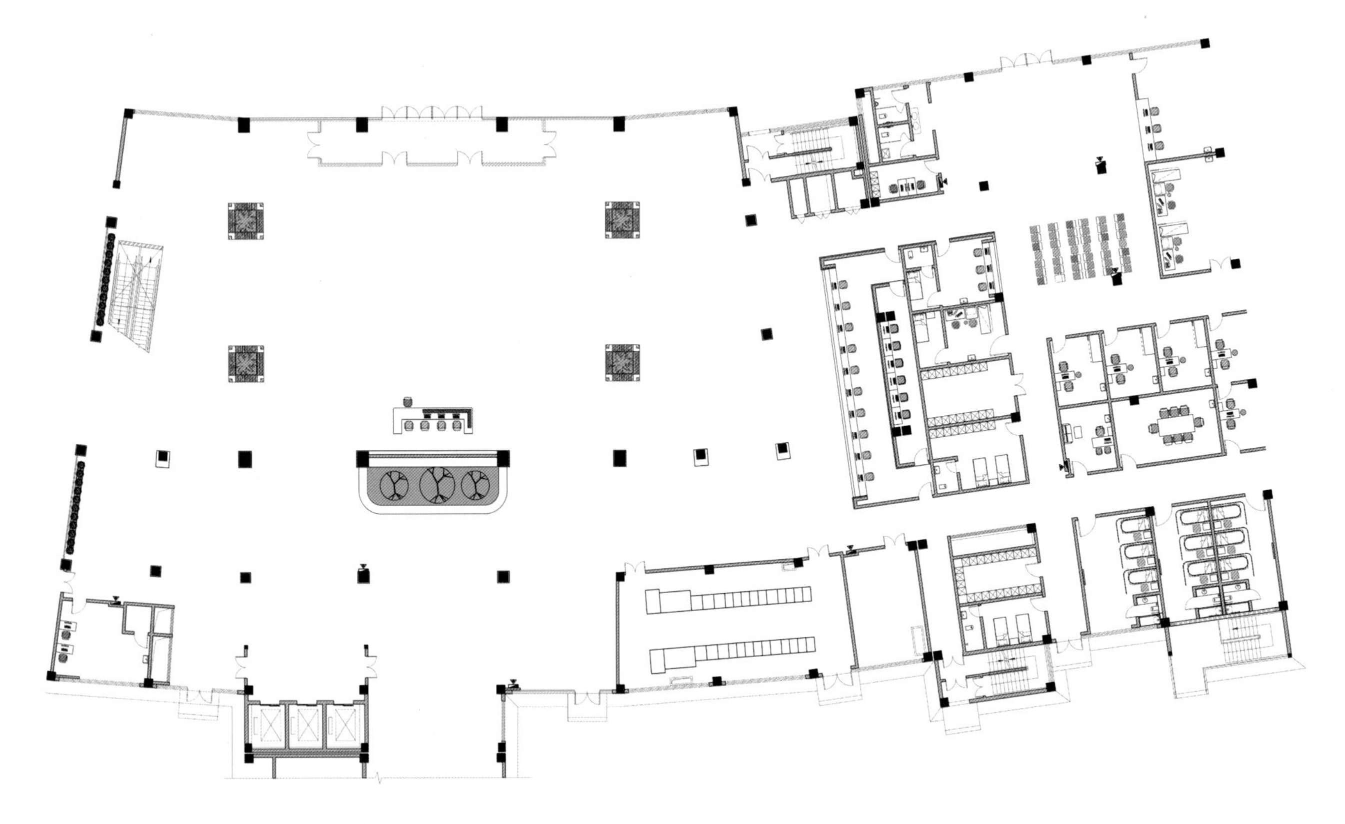

明德厚医
求实拓新

明德厚医
求实拓新

医学生誓言
健康所系，性命相托。
当我步入神圣医学学府的时刻，谨庄
严宣誓：
我志愿献身医学，热爱祖国，忠于人
民，恪守医德，尊师守纪，刻苦钻研，孜
孜不倦，精益求精，全面发展。
我决心竭尽全力除人类之病痛，助健康之
完美，维护医术的圣洁和荣誉，救死扶伤
，不辞艰辛，执着追求，为祖国医药卫生
事业的发展和人类身心健康奋斗终生。
THE OATH OF A MEDICAL STUDENT
Health related, life entrusted.
The moment I step into the hallowed medical
institution, I pledge solemnly
I will volunteer myself to medicine with
love for my motherland and loyalty to the
people.
I will scrupulously abide by the medical
ethics ,respect my teachers and discipline
myself.
I will strive diligently for the perfection
of technology and for all-round development of
myself.
I am determined to strive diligently to
eliminate man' s suffering, enhance man' s
health conditions and uphold the chasteness and
honor of medicine.
I will heal the wounded and rescue the dying
,regardless of the hardships.
will always be in earnest pursuit of better
achievement .I will work all my life for the
development of the nation' s medical enterprise
as well as mankind' s physical and mental
health.

襄阳市襄州区人民医院，建筑面积 160000m^2，目前开放床位 938 张。为了避使功能区更加合理化，在技术上采用系统集成的装饰手法，从而达到装饰造价的节约目标。

Xiangyang Xiangzhou District People's Hospital with a building area of 160,000 square meters opens 938 beds at present. Technically, systematic and composite methods of decoration are adopted so that the decoration costs are reduced.

外科门诊
Surgical Clinic

护士站 Nurse Station
体检中心

儿科护士站

DEPART ENT
OF NEO TOLO

PACE OF
LIFE MEDICINE
生命医学的脚步
2013
1936
NURAE STATION

2F
电梯厅
NURAE STATION

山东省枣庄市妇幼保健院

Maternal & Child Health Care of Zaozhuang

为了缓解枣庄全市妇女儿童看病难、住院难的问题，枣庄妇幼保健院进行了扩建，是市级非营利性医疗保健机构，属三级甲等妇幼保健院。为了贴合此次项目的妇幼客群，以石榴花以及剥茧成蝶等元素运用其中，古人称石榴“千房同膜，千子如一”，寓意多子多福，且石榴花为枣庄市花，贴切主题的同时又有地理缘由。

In order to solve the problems of local medical service, Maternal & Child Health Care of Zaozhuang has been relocated and reconstructed. It is a city-level non-profitable health care institute, one of the class-A maternal and health care hospitals. Taking the target customers into account, the elements of pomegranate blossom and the evolution of pupa to butterfly are employed in its design. Because of its honeycomb structure, the pomegranate has long been considered the symbol of fertility and prosperity. Besides, the selection of pomegranate blossom, the city flower, fits the design theme and realizes localization.

VECTOR

为了解决儿童在输液时需要家属看护的问题，设计中采用围合的输液椅，便于家属在不受干扰的同时集中注意力照顾儿童。

The innovation of infusion chair in the design is convenient for the parents to care for their children.

A3
A2
B1
C3
C2
C1

Dentistry

卫生间的洗手池以高低台的形式设计，以便儿童使用，卡通元素亦能营造儿童空间氛围。

The design of double-level sink is convenient for children and the cartoon image ornaments also create children's space atmosphere.

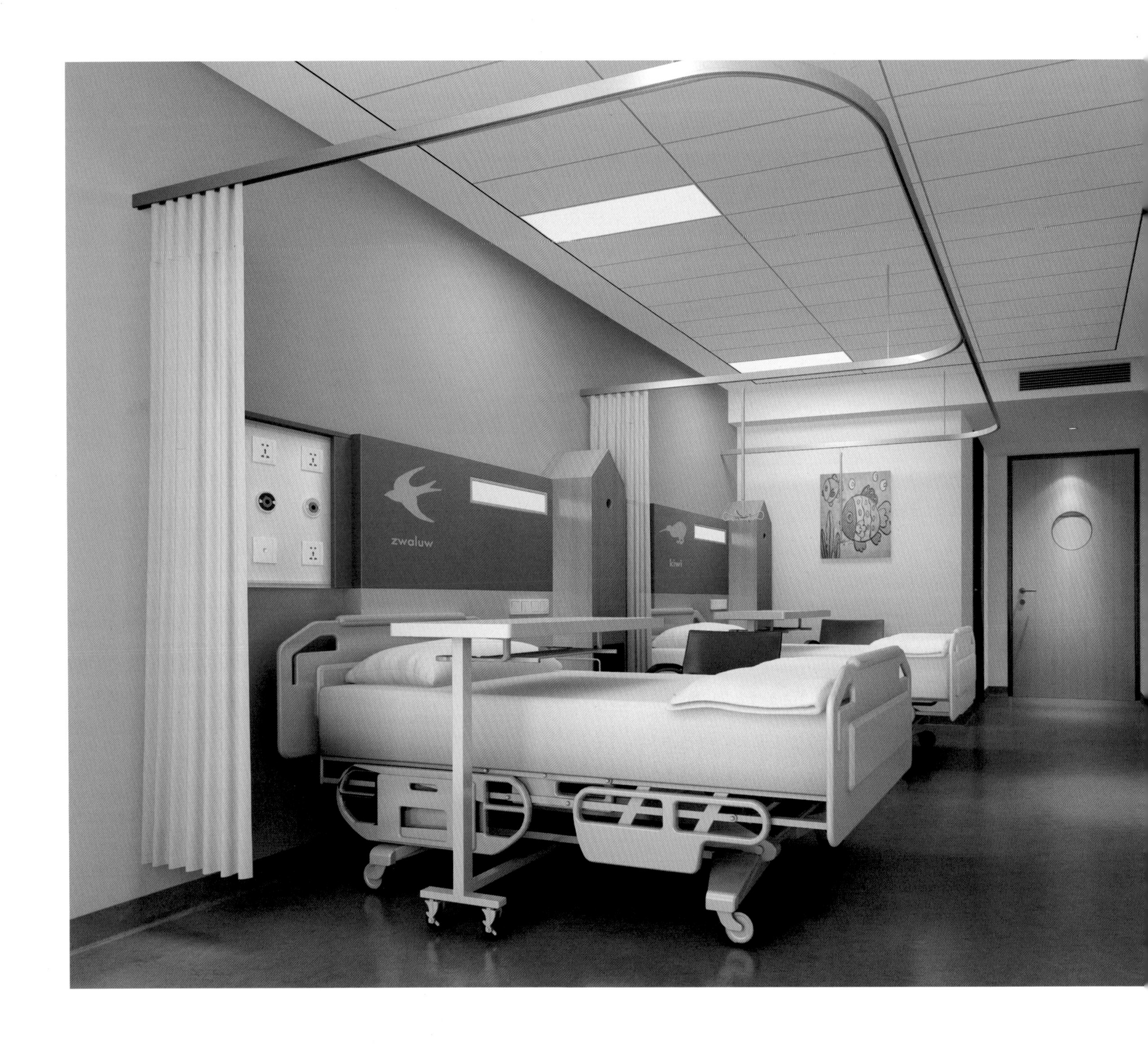

为了解决坐便器接地打胶带来的发霉发黑的问题，设计中使用了离地式坐便器。淋浴区采用不锈钢水篦子，不仅能快速排水，亦能防止病患被拦水台阶绊倒。同时还有淋浴座椅、紧急呼叫按钮以及上翻式扶手，将人性化植入到每个细节中。

The toilets in the hospital are installed above the ground to avoid mold or blackening. Stainless perforated strainer is used in the bathroom to avoid tripping on the boundary of shower tray. Humanity design concept is implanted, revealing in every detail from shower chair, emergency alarm to reversible rail.

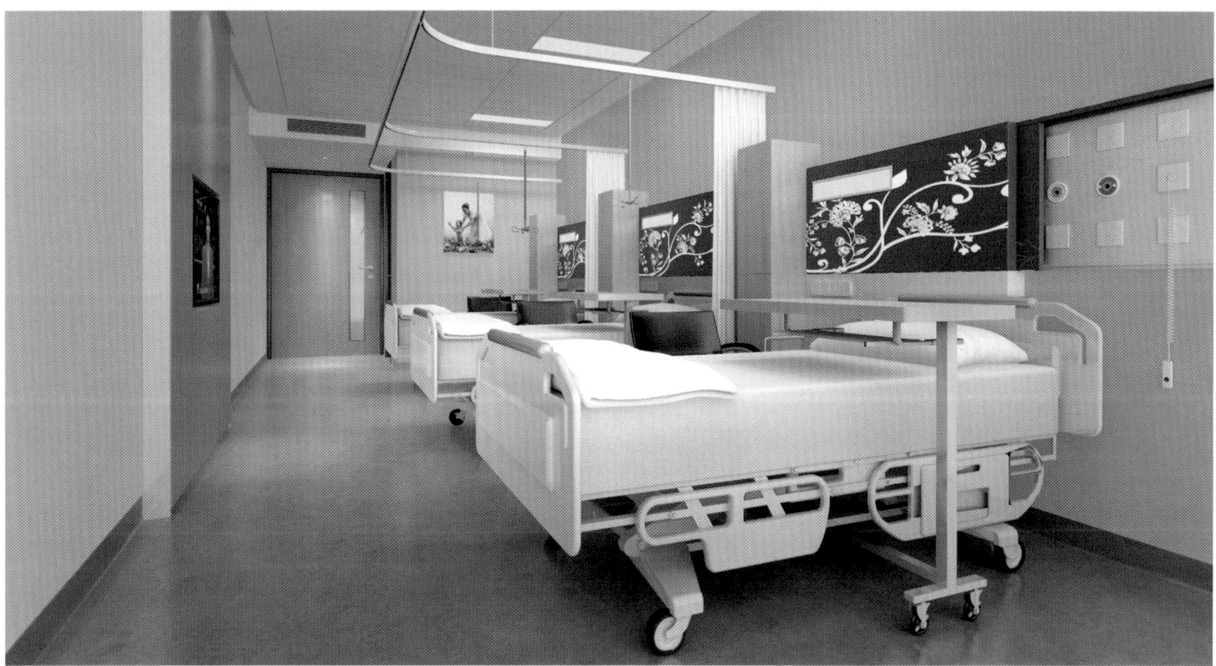

枣庄妇幼保健院，建筑面积 120000m²，预计设置床位 700 张。在灯光设计中，采用不同于其他医院的圆环灯设计，既有对水元素的应用，亦有软化空间氛围的作用，给妇幼人群一个全新的就诊环境。

Maternal & Child Health Care of Zaozhuang with a building area of 120,000 square meters is scheduled to open 700 beds. In its unusual lighting design, lamps are installed in circle not only to apply the water element but also to lighten the atmosphere, exposing the women and children to a brand new medical environment.

SEA
WORLD
奇幻海底
护士站 NURSE STATION
NURSE STATION

山东省聊城市东昌府区妇幼保健院

Liaocheng Dongchangfu District Women & Children Health Care Hospital

古有聊河，京杭大运河的商业文明与黄河的农业文明在此交汇，聊城因此而得名，素有“北水城”之称。设计中运用了大量的水元素，在空间中将水景与二级导视相结合，形成生命活水之寓意。此外提取水的字形元素，将等候椅设计成波浪的韵律。与水相连，形成灵动的空间氛围。

As the name given, Liaocheng, the city where the commercial culture of the Great Channel, the Liao River and the agricultural culture of the Yellow River met in the history, has enjoyed the reputation of "North Waters City". Therefore, a great many water elements are employed in the design to mix the waterscape with the second-level navigation signs, suggesting the vigor of lively water. Besides, inspired by the Chinese character of water, lines of waiting chairs are visually designed in rhythm with waves, creating a lively atmosphere.

1F
Information
Register & Pharmacy
Consulting Room

总导台处，水经由导视装置流向水景，另有鱼缸等装置，增加自然的人文气息。

At Reception, the interior water steered by visual devices streams to the waterscape. In addition, the fish tank and other installations add a flavor of humanism to nature.

NURSE STATION

PERINATAL CENTER
围产医学中心

Empowering

护士站
NURSE STATION
Nurses station 护士台

聊城东昌府区妇幼保健院建筑面积 57000m²，设有床位 800 余张。采用不冗余式的装饰材料，主题色调以灰白、木黄为主，以少量亮色点缀，带有极强的地方特色，采用正压式空气流通系统，完美地解决了医院空气不流通带来的污染隐患。

Liaocheng Dongchang District Women & Children Health Care Hospital with a building area of 57,000 square meters opens more than 800 beds. Having no redundant materials, the hospital whose tones are mainly grayish white and wooden yellow with a few bright ornaments, possesses noticeable local features. The positive pressure air circulation system is adopted to avoid the potential danger caused by inadequate ventilation.

江苏省淮安市妇幼保健院

Huai'an Women & Children Health Care Hospital, Jiangsu

有着 58 年历史的淮安妇幼保健院，坐落于风景秀丽的清晏园畔，是淮安地区唯一的三级甲等妇幼保健机构。设计的元素从 DNA 等医学分子、医护工作状态、破茧成蝶中提取。针对女性客群，采用玫红色作为人流量大区域的背景色，使空间更具柔美气质。

Huai'an Women & Children Health Care Hospital, located alongside the scenic Qingyan Garden, is the only class-A women and children health care institute in Huai'an. The design elements include DNA, medical molecules, working scenes and the evolution of pupa to butterfly. Aimed at women customers, the color of red rose is the background color of crowed areas, making the interior space filled with feminine elegance.

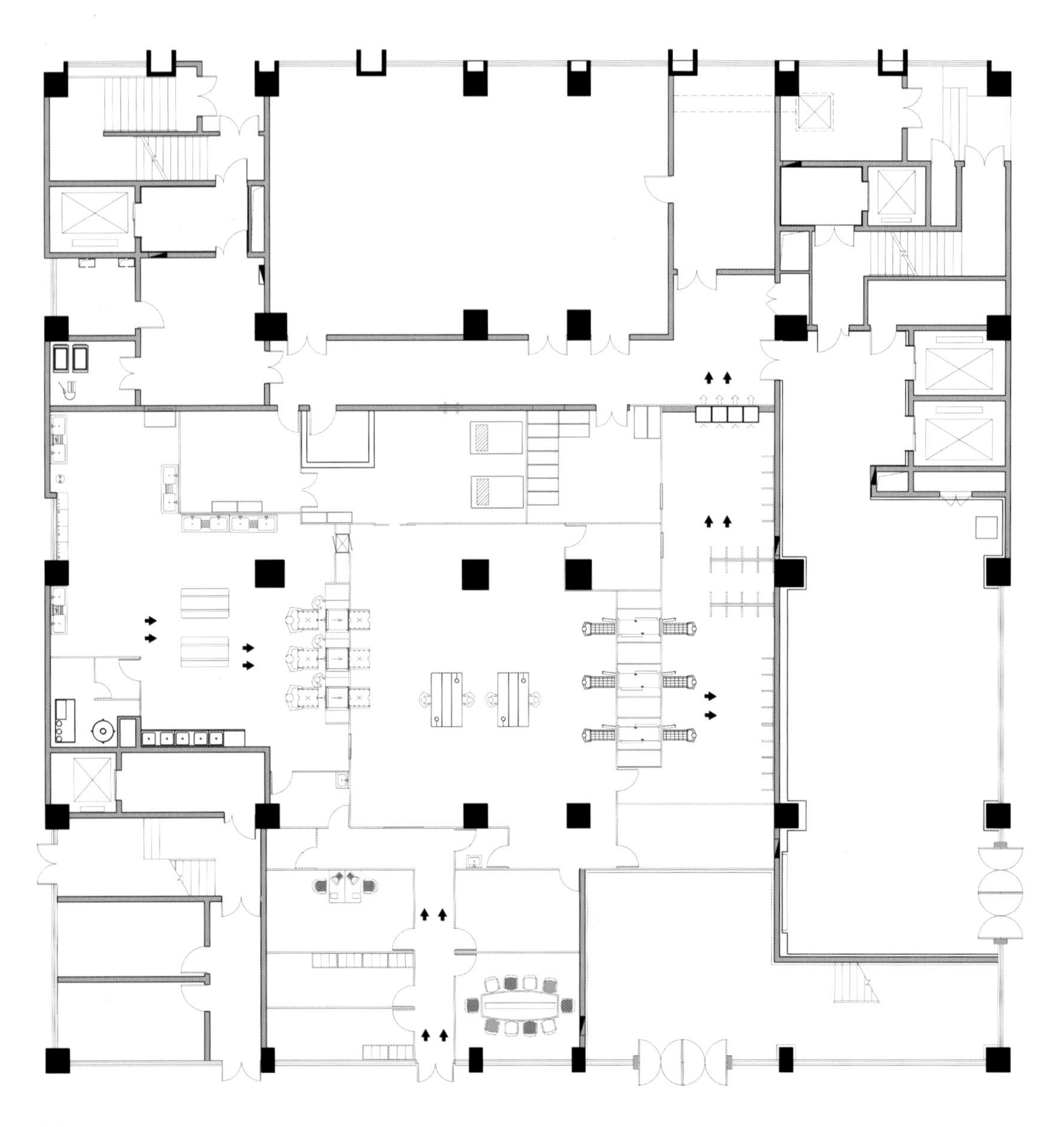

护士站
Nurse Station

2013
仁爱 敬业 博学

PACE OF
LIFE MEDICINE
生命医学的脚步
2010'
1958'
2013'
2009'
information/press

服务台采用圆弧面设计，可有效防止棱角伤害。顶面采用可拆卸穿孔材料，解决了管线升级问题并能有效降噪。

The edge of the service counter is curved so as to escape potential danger. On the top, perforated materials are detachable for the purpose of line upgrade and noises reduction.

护士站
Nurse Station
PACE OF
LIFE MEDICINE
生命医学的脚步
1958'
2009'

淮安妇幼保健院，建筑面积 50100m²，开放床位 600 张。设计主色调为灰白、草绿及玫红。住院部门厅开阔，所有的事情都在这里发生。门厅外部采用全玻璃拼装，自然采光性能得到有效保障。

Huai'an Women & Children Health Care Hospital with a building area of 50,100 square meters opens 600 beds. The main tones of the hospital are grayish white, grass green and red rose colors. The out-patient lobby is spacious and its exterior is assembled by glass, which ensures the natural lighting performance.

山东省聊城市东昌府人民医院

Liaocheng Dongchangfu District people's Hospital

聊城市东昌府人民医院始建于 1949 年，是一所集医疗、预防、保健、 康复、教学和科研为一体的二级甲等综合性医院，是东昌府区医疗技术指导中心和急救中心，市、区两级医保、新农合定点医院。本次设计为新院建设，建筑面积 143021 平米。

“蝉噪林愈静，鸟鸣山更幽”本案以森林为主题，给患者营造了一个充满自然气息的空间，游鱼、竹林、花草、鸟鸣将人们引入大自然。

Liaocheng Dongchangfu District people's Hospital, founded in1949, is a general hospital of health care, prevention, rehabilitation, teaching and scientific research. This rank Two, grade A hospital is also the medical guidance center and emergency center of Dongchangfu district. Health care is both at the municipal and district levels and it is also the new type of rural cooperative medical hospital.

The design is for this new hospital and the building area is about 143,021 square meters. "The forest is more peaceful while cicadas are chirping and the mountain is more secluded while the birds are singing." This case is the theme of forests and creates a space full of natural characters such as bamboo, flowers, birds and fish.

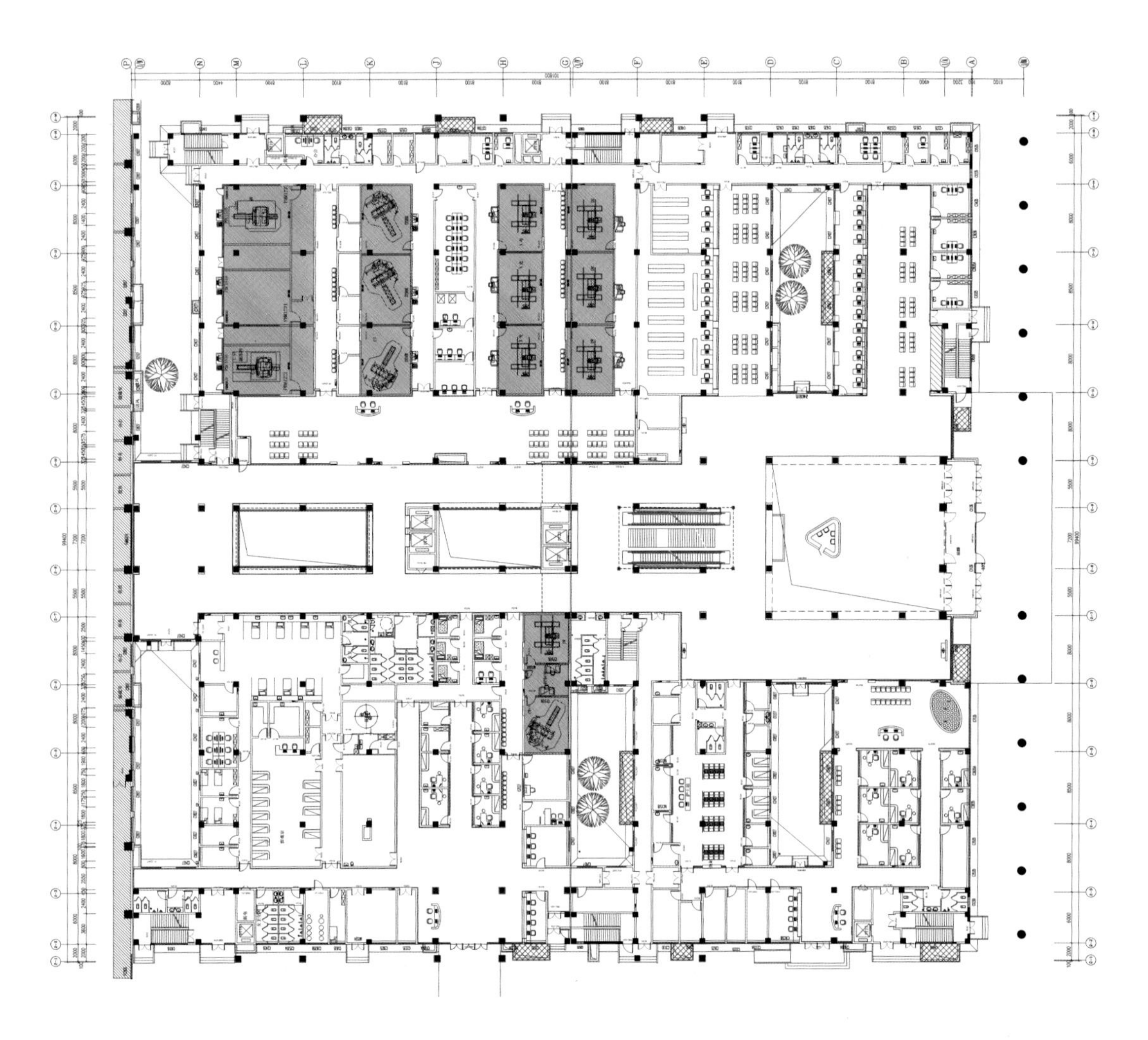

厚德 精术 严谨

A
挂号收费
REGISTRATION FEE
A1 挂号收费
REGISTRATION FEE
B

Consulting Room

10
检验科

NURSE STATION

NURSE STRTION

护士站 Nurse Station
护士站
护士站 | NURSE STATION
NURSING
STAITON

05
Cardiology
Department
护士站

江苏省常州市第四人民医院（肿瘤医院）
Changzhou Fourth People's Hospital (Tumor Diseases Hospital)

常州别称“龙城”，有“三吴重镇，八邑名都”之称。拥有60多年历史的常州第四人民医院，以肿瘤防治及急诊创伤为主要特色。空间设计中，将常州的市树广玉兰、竹子、DNA融入其中，广玉兰因生命力强，被运用于服务台立面、主立面墙、家具等处，将DNA的自然形态进行提炼，以双螺旋的造型路灯彰显医院的专业底蕴。

Changzhou is also known as "Dragon City" for its important position in ancient times. With a history of 60 years or more, Changzhou Fourth People's Hospital specializes in tumor prevention and control and the emergency trauma treatment. In its interior design, magnolia, bamboo and DNA are employed. Because of its indomitable vitality, the element of magnolia is applied on the reception desk front, the main vertical wall and the furniture. Inspired by the natural form of DNA, the double-spiral shaped street lamps manifest professional background of the hospital.

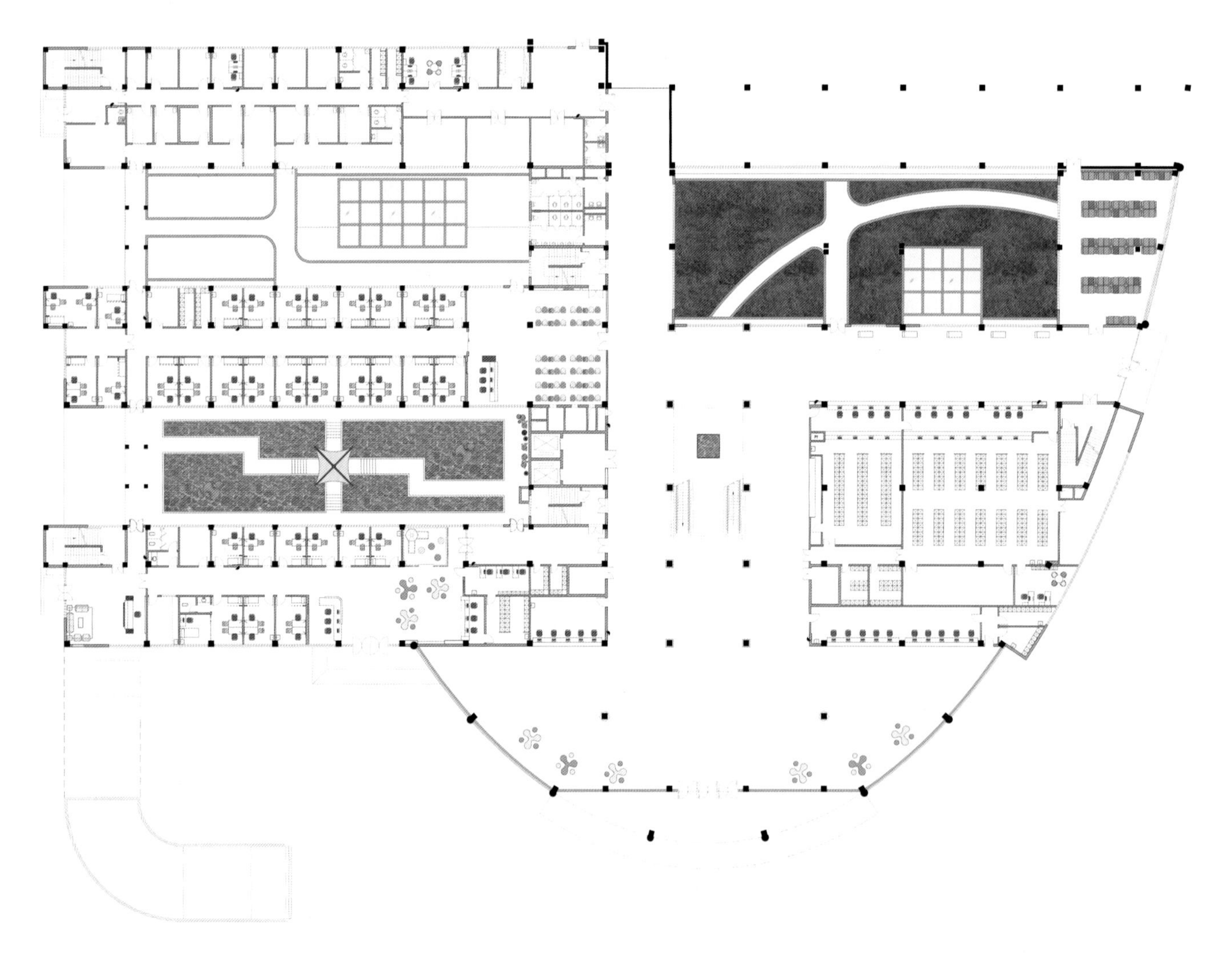

大厅内使用竹子的艺术变形设计圆形包柱，将自然气息融入其中。

Circular pillars in the lobby are designed in refined bamboo form, which infuses a flavor of nature into the lobby.

药房 pharmacy
入院处 in patient admissio
收费处 Csahier
1F
电梯
Elevator

电梯
Elevator

护 士 站
Nurse Station

6F

洗手间
Toilet
1F
中西药房
Pharmacy
1取药
2取药

出入院登记 Registry
1收费
2收费

休闲区Leisure areas
超市Supermarket

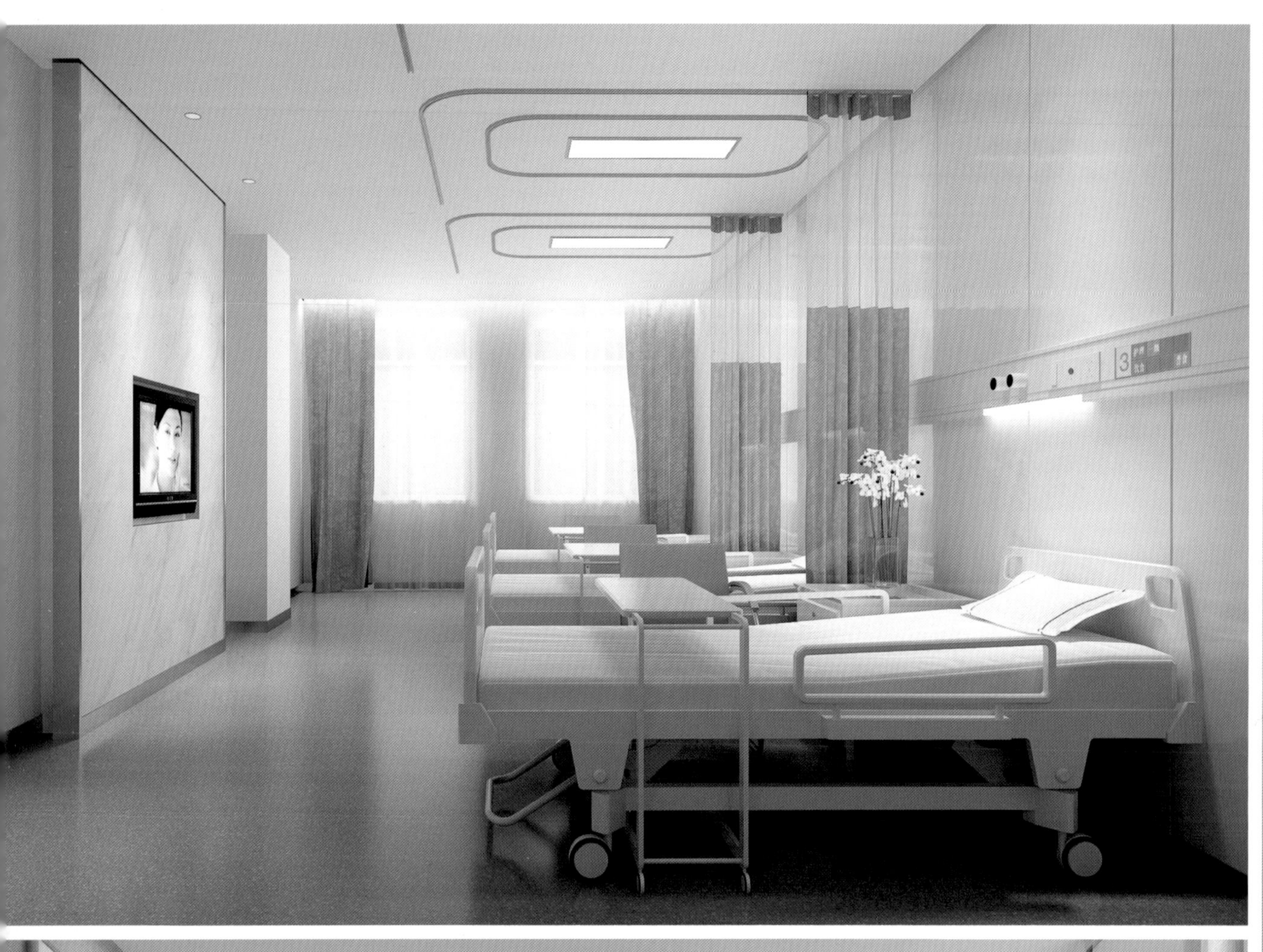

NURSE STATION
NURSE STATION/护士站

陕西省西北妇女儿童医院

Shaanxi Province Northwest Women & Children Hospital

西北妇女儿童医院是陕西省卫生计生委直属的三级甲等专科医院。儿童对于蓝天、白云、大海的喜爱是天生的，在西北妇女儿童医院的设计中，加入了这些设计元素，使得整体空间氛围趋于活跃。以简洁的材料与色彩打造愉悦的就医氛围。妇科门诊处则以西安石榴为特色元素，与蝴蝶简化成图形装点空间，玫红色成为主体色，拉近女性与医院的距离。

Shaanxi Province Northwest Women & Children Hospital, directly subordinated to the National Health and Family Planning Commission of Shaanxi Province, is a class-A specialized hospital. The child interest in sky, cloud and the sea is born in nature, and therefore, these elements are included in the design, which makes the overall atmosphere alive. Simple materials and colors create a pleasant medical treatment atmosphere. The gynecology clinic, for example, pomegranate, the remarkable feature of Xi'an, is used in accompany with simplified butterfly-shaped ornaments, and the tone of red rose color narrows the distance between women patients and the hospital.

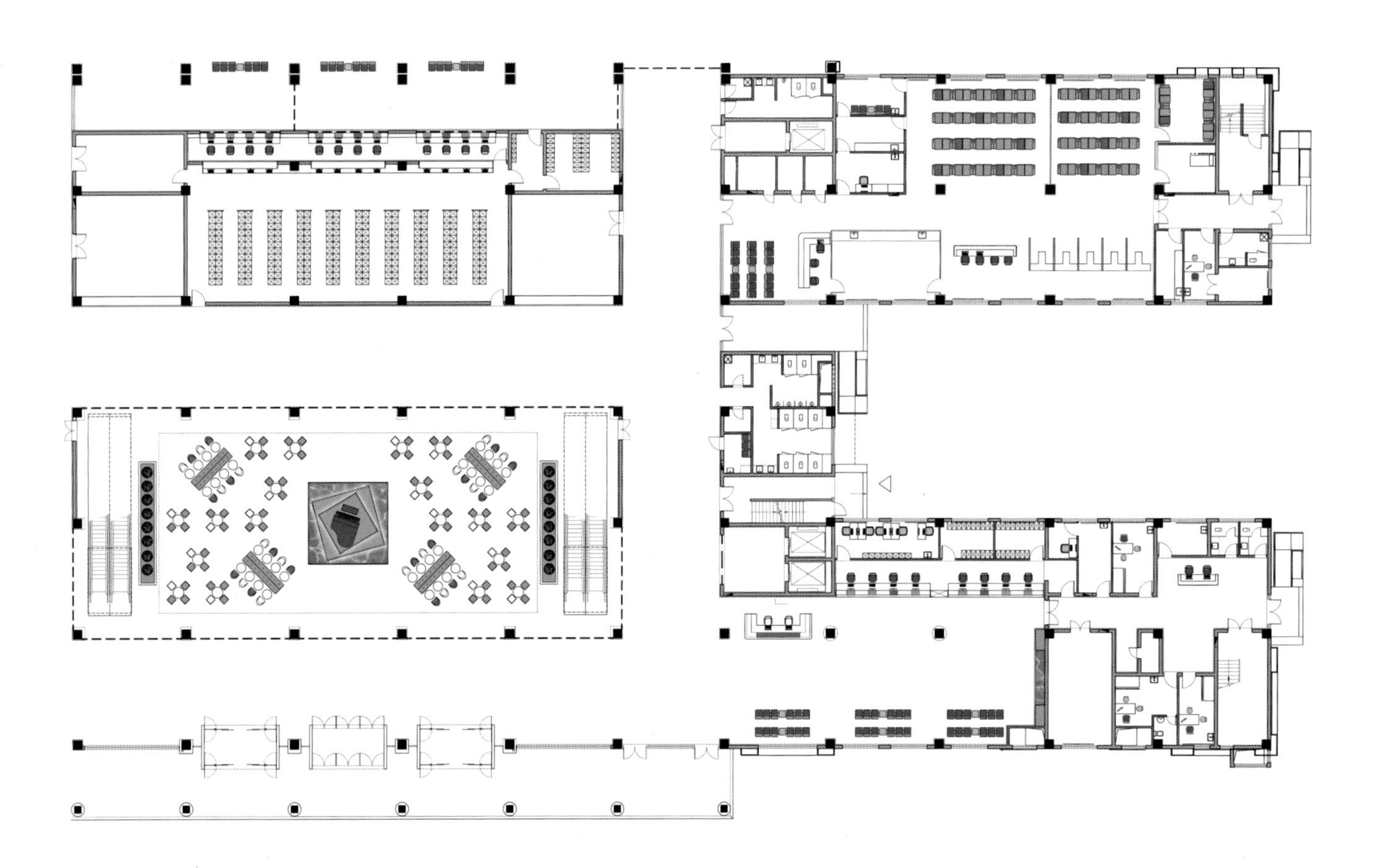

儿
科
6区
OFFICE

芯创领·新跨越

高低台的设计使得挂号收费服务台能够同时满足成人、儿童以及残疾人的使用需求，更具人性化。鱼缸采用圆形包裹，更显趣味性。

The double-level counter design of Register and Charge Window is more humanized, allowing easy access to the adults, children and the handicapped. The fish tanks are all wrapped in circular shape, adding more delight and interest to the interior space.

西药房
Dispensary
取药
取药
1F

检验
检验

护士站
Nurse Station
1区

2区

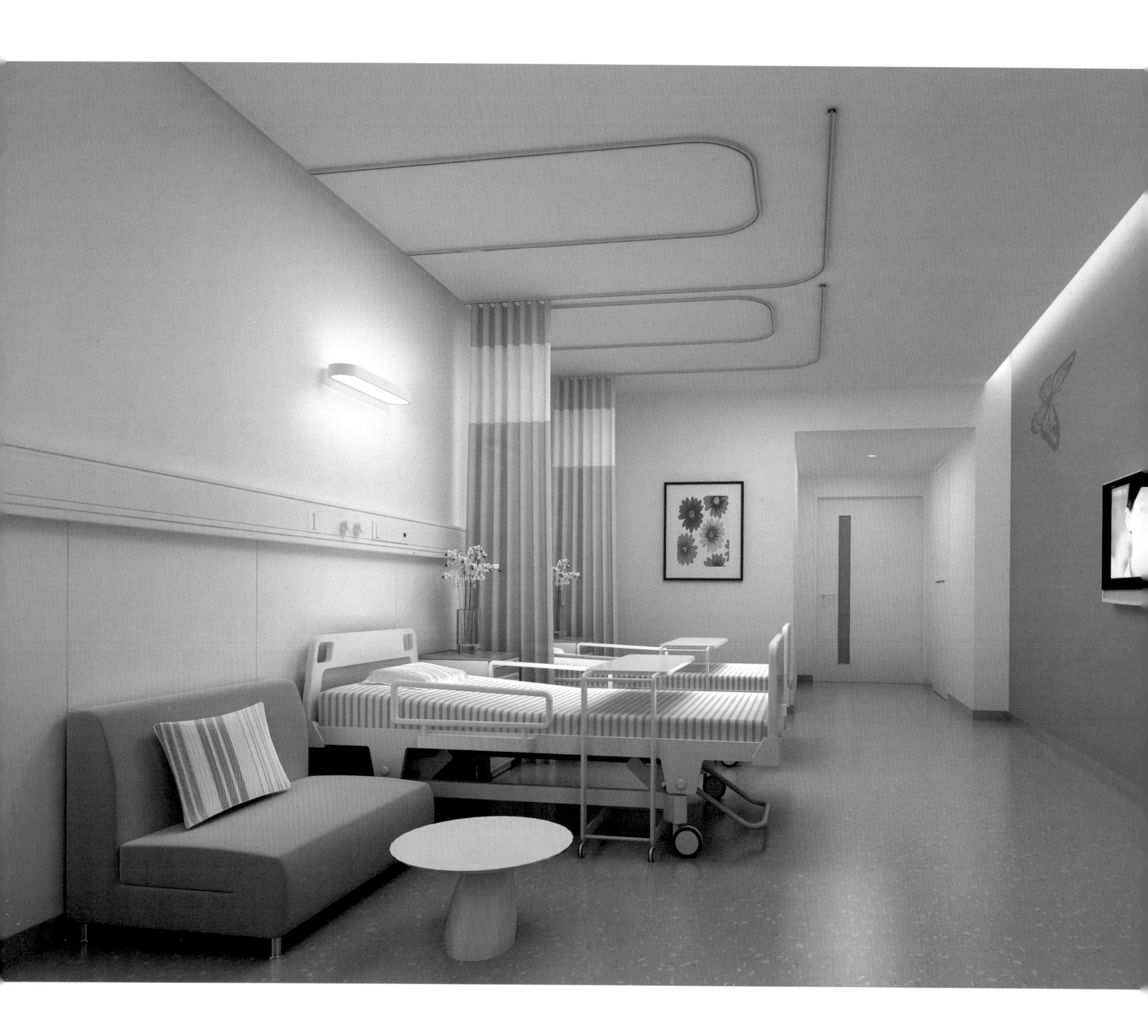

西北妇女儿童医院，建筑面积 53300m^2，设有床位 2000 张。符合儿童审美的颜色搭配是该空间的特色，柠檬草饮料般的绿色、如美味的提神糖果般的蓝色、如橘子皮般的橙色，再加以动物卡通造型于其中，打造颇具生动特色的儿童就医空间。

Northwest Women & Children Hospital with a building area of 53,300 square meters opens 2,000 beds. Its spatial feature is the good match of colors which is in agreement with children's judgment on beauty, like the colors of lemon green, refreshing blue and marmalade orange. The match coupled with cartoon images creates a quite lively hospital environment for children.

护士站
Nurse Station

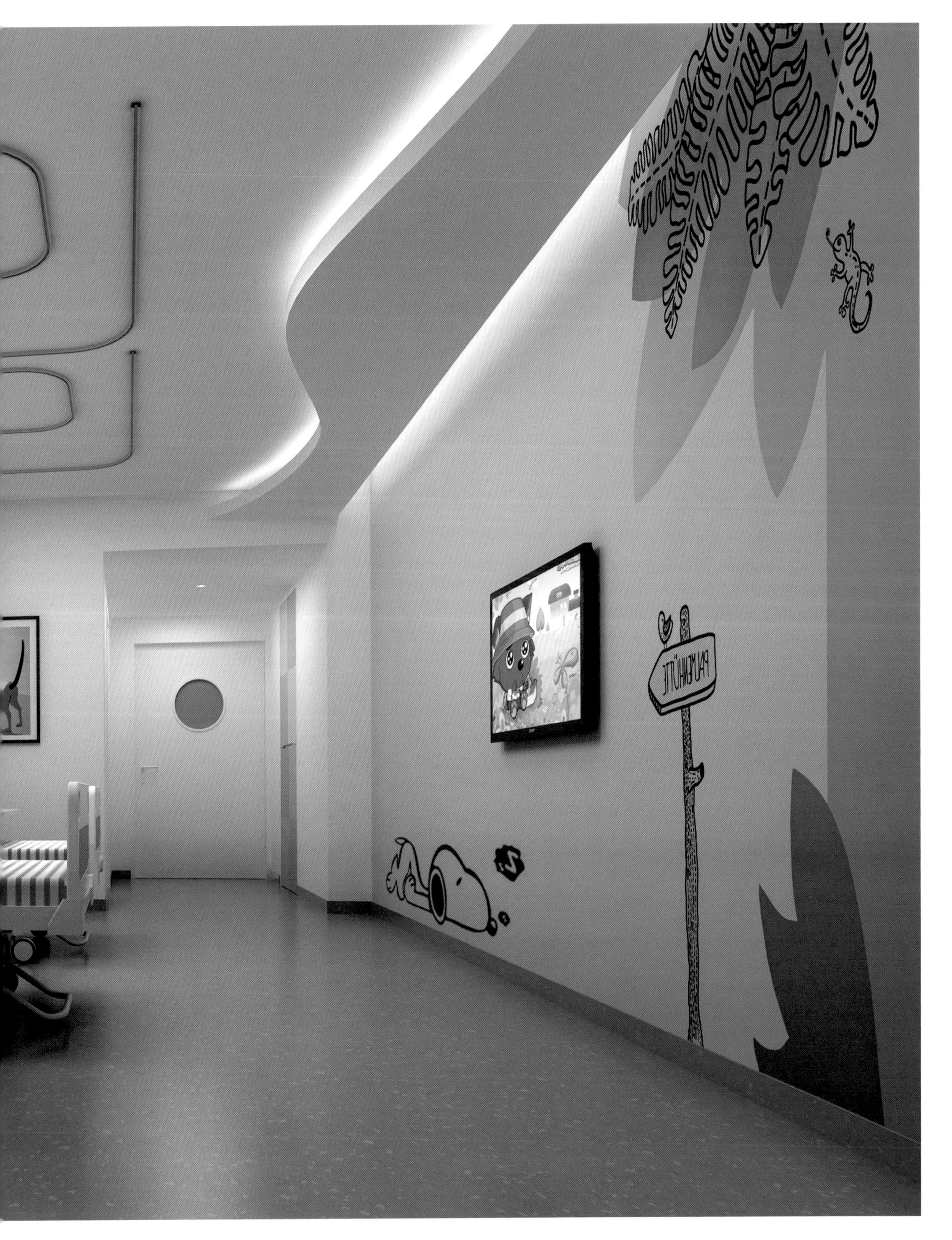

江苏省南京河西地区老年康复疗养中心

Nanjing Hexi Rehabilitation & Rest Center for Seniors

南京河西地区老年康复疗养中心（河西莲花社区服务中心）位于南京市建邺区黄河路以南、莲花村路以东，是一所多功能社区服务中心。随着我国人口“老龄化”的现象越来越明显，康复疗养、老年公寓等空间的设计需求也在日益提高。将中式花格与“福、寿”琉璃艺术品搭配，并且空间顶部运用了南京市花——梅花的元素，寓意对老人的诚挚祝福。整体风格简约自然，益于老年人静心休养。

河西莲花社区服务中心，建筑面积 52000m²，是一所多功能社区服务中心，其中派出所 3121m²，老年社区医疗康复中心 1500m²，老年公寓 17546m²。其中老年人使用部分设计主色调以灰、白、黄为主，暖色系材料使得使用者无时不刻都能有积极的心理暗示。在起居室以及卫生间，采用了多项细节技术，让老年生活中的种种潜在风险得以解决，从空间设计上让老年人安居于此。

Nanjing Hexi Rehabilitation &Rest Center for Seniors (Hexi Lianhua Community Service Center), located in Jianye district of Nanjing, south of Yellow River Road and east of Lianhua Village, is a multi-functional community service center. As the aging problem becomes increasingly apparent, the need for rehabilitation and rest and senior housing apartments are on the rise. Therefore, the Chinese style window frames matching colored glaze ornaments embedded with Chinese characters “ 福 ” and “ 寿 ”(which means “blessing” and “longevity”), together with the element of plum blossom, the city flower of Nanjing, on the top express sincere wish to the seniors. The overall style is natural and simplistic, preferable to meditation and rest.

As a multi-functional service center, Hexi Lianhua Community Service Center has a building area of 52,000 square meters, of which the local police station covers 3,121 square meters, the medical rehabilitation center 1,500 square meters and the senior housing apartments 17,546 square meters. The main tones of the design for seniors are gray, white and yellow colors, and materials in warm tones constantly offer the seniors positive psychological hints. In the living room and toilet design, many details have been taken into account so that potential dangers are minimized, which ensures the seniors a happy later life.

和氣致祥

活动中心设计通过玻璃窗将自然景观引入室内，内外呼应。地面采用橡胶地板，减少老人摔倒伤害。顶部采用吸音材料，有效降低运动带来的噪音。

The windows of the fitness center render everything in front more graphically defined. The plastic flooring reduces harm to the seniors in case they slip. Moreover, soundproofing materials are used in the ceiling design, effectively reducing the sound when the seniors are exercising.

接待室采用木纹色墙面和米白色家具搭配，梅花元素的吊灯与地毯相互呼应，古典雅致亦不失现代感。

The walls of the reception room are painted wood grain color and the furniture there is creamy white. The plum flower petal shaped ceiling lamp corresponds with the carpet, classic and modern.

电梯轿厢顶面采用无暗区灯光，减少密闭空间带来的眩晕和不适感。并设置了安全镜方便轮椅使用者进出时观察后方。

Top lighting of elevator cabs is free of dead corner, which reduces dazzling or uncomfortable feelings. Besides, rear windows are installed for the convenience of wheelchair users.

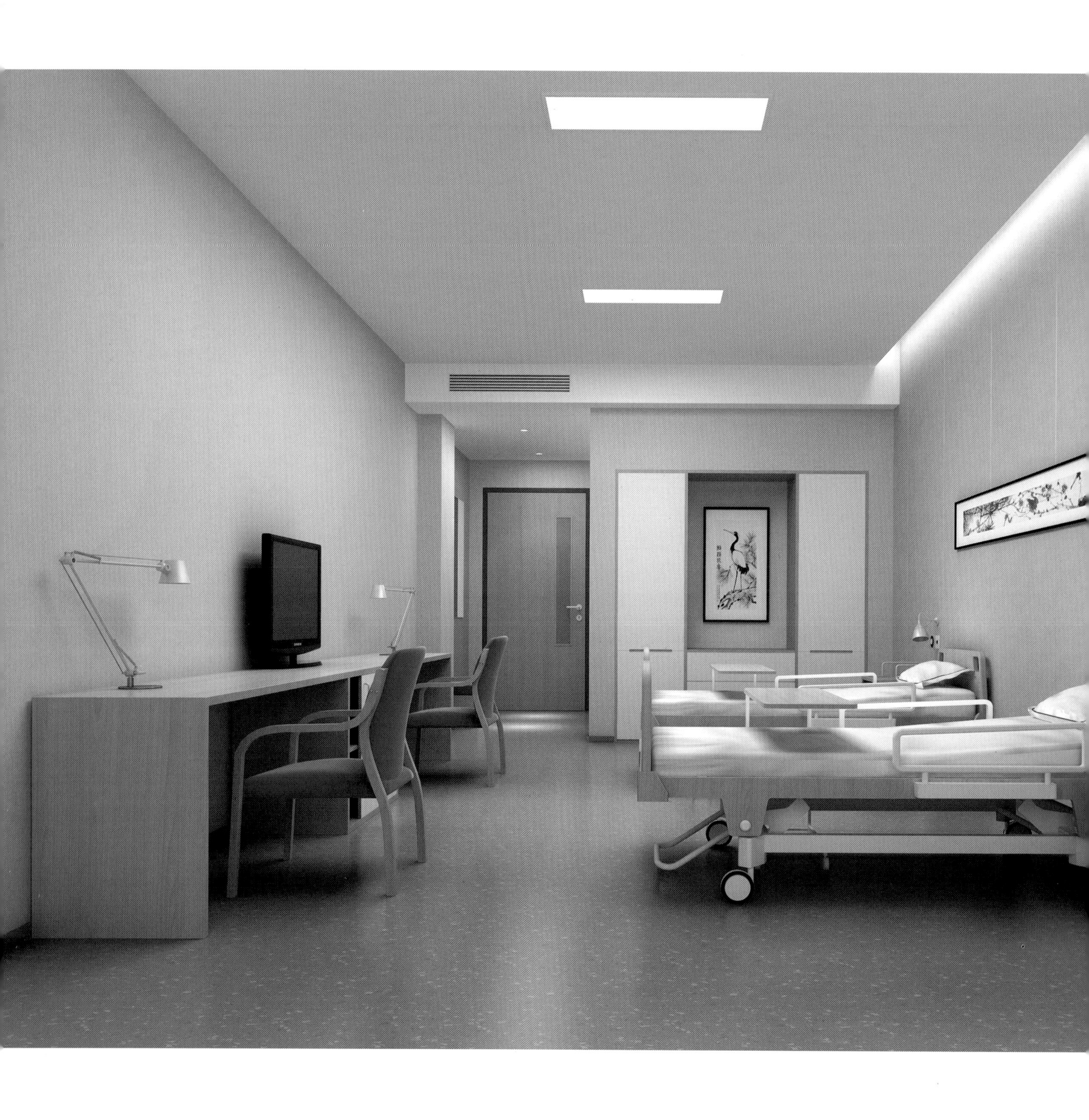

邯郸市第七医院（妇产儿童医院）

Handan No.7 Hospital (Women and Children's Hospital)

邯郸市第七医院是国家二级甲等综合医院。始建于 1947 年，位于邯郸市雪驰路 33 号，是一所集医疗、预防、保健、康复、科研、教学于一体的大型综合医院，是全市及周边地区人民群众的医疗、保健中心。医院分为老院区、东院区、新院区迁建项目。此次设计为东院区妇产儿童医院，主要为妇产门诊、医技、月子中心、体检中心，建筑面积 20000m^2。

Handan No.7 Hospital is a comprehensive national-level class-B hospital. Founded in 1947, the hospital is located at No.33 Xuechi Road with an integration of medical care, prevention and control, health care, rehabilitation, scientific research and education. It also serves as the medical and health care center for Handan city and its neighborhood. The hospital consists of the old campus, the eastern campus and a relocated campus currently under construction. The case is the design of Women and Children's Hospital in the eastern campus with a construction area of 20,000 square meters, including Obstetrics and Gynecology Clinic, Medical Technology Department, Confinement Center and Physical Examination Center.

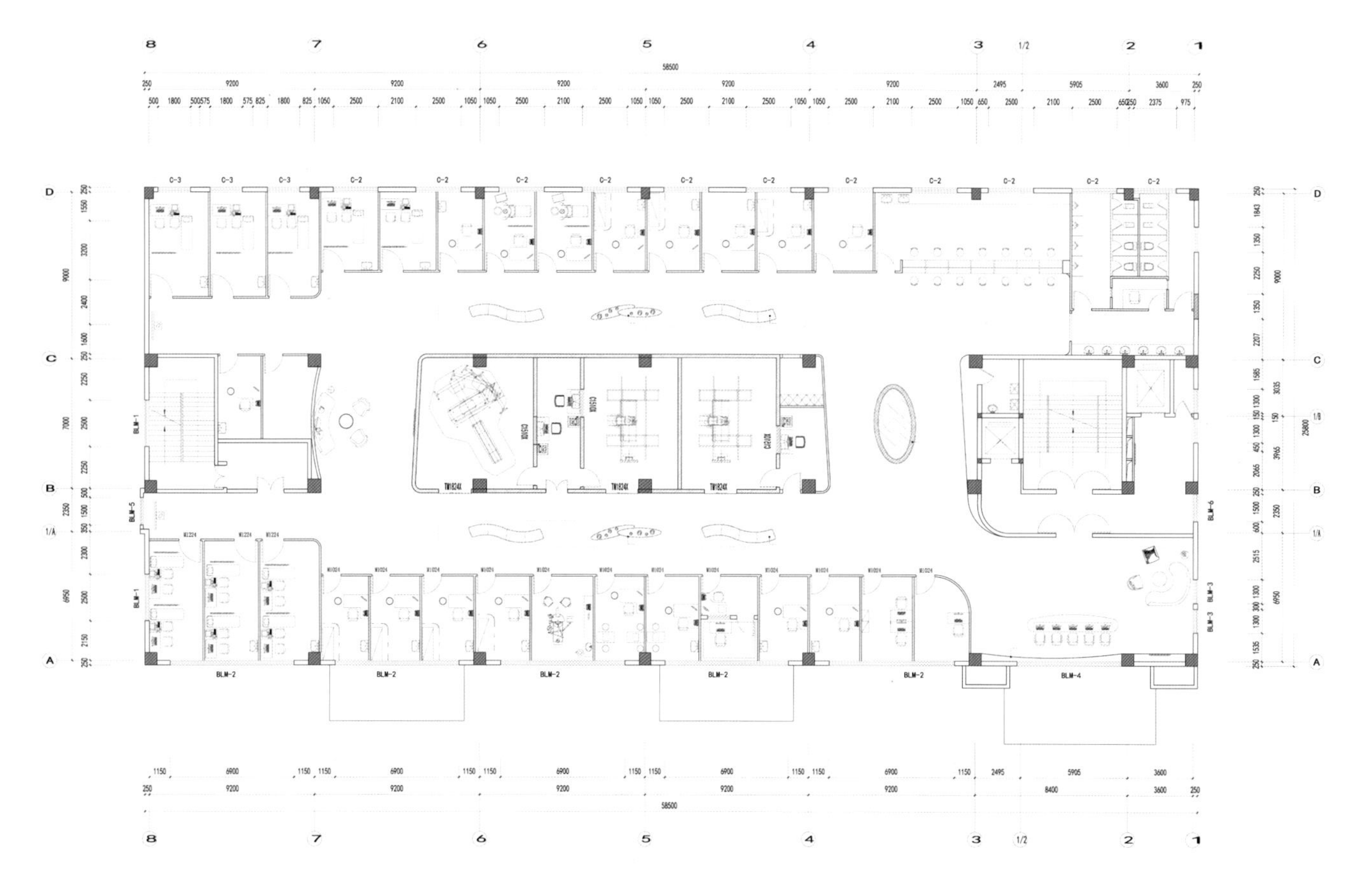

1F
Information desk

建筑是凝固的音乐，音乐是流动的建筑。黑格尔曾这样提示音乐与建筑的关系：“音乐和建筑最相近，因为像建筑一样，音乐把它的创造放在比例和结构上。”建筑是一种以形式为主的造型艺术，它能激起同听音乐相近的情感反应。

As the architecture is a piece of fixed music, the music is a flowing architecture. Hegel once explained the association between architecture and music. In his eyes, the music is the most similar form of art to the architecture for their same emphasis on proportion and structure. Following the philosophy, we consider architecture "plastic arts", which similar to the music, can also provoke emotional reaction.

04
检验抽血
REGISTRATION FEE

Medical Center

Maternity hotels

本案设计以流动的建筑为主题，运用起承转合的结构形式将曲线融入室内空间，从邯郸当地自然景观、特色文化、市花、市树中提取自然元素，通过艺术的融合与提炼，用简化、叠加、重合手法，将自然元素延伸并融入到室内空间，力求塑造一个具有特色、时尚、现代化医疗空间。

This design takes the theme of flowing architecture, employing curves in the interior, achieving the transitional effect of the layout. The refined natural elements of the landscape, distinctive culture, and the city flower and tree of Handan are mixed with fine arts. By simplification, overlapping and superposition, these natural elements extend and fuse into the interior, managing to form a distinctive, fashionable and modern medical space.

套间VIP
502
501

6
产科病房
maternity ward

南通大学附属医院体检中心

Medical Examination Center, Affiliated Hospital of Nantong University

南通大学附属医院由清末状元张謇先生创建于 1911 年。1994 年被卫生部首批评定为三级甲等综合性医院，1998 年被确定为国际紧急救援中心网络医院。

Affiliated Hospital of Nantong University was founded in 1911 by Zhang Yu, an imperial examination champion in late Qing dynasty. In 1994, it was graded as one of the first group of class-A hospitals by the Ministry of Health. In 1998, it is identified as international emergency aid center online hospital.

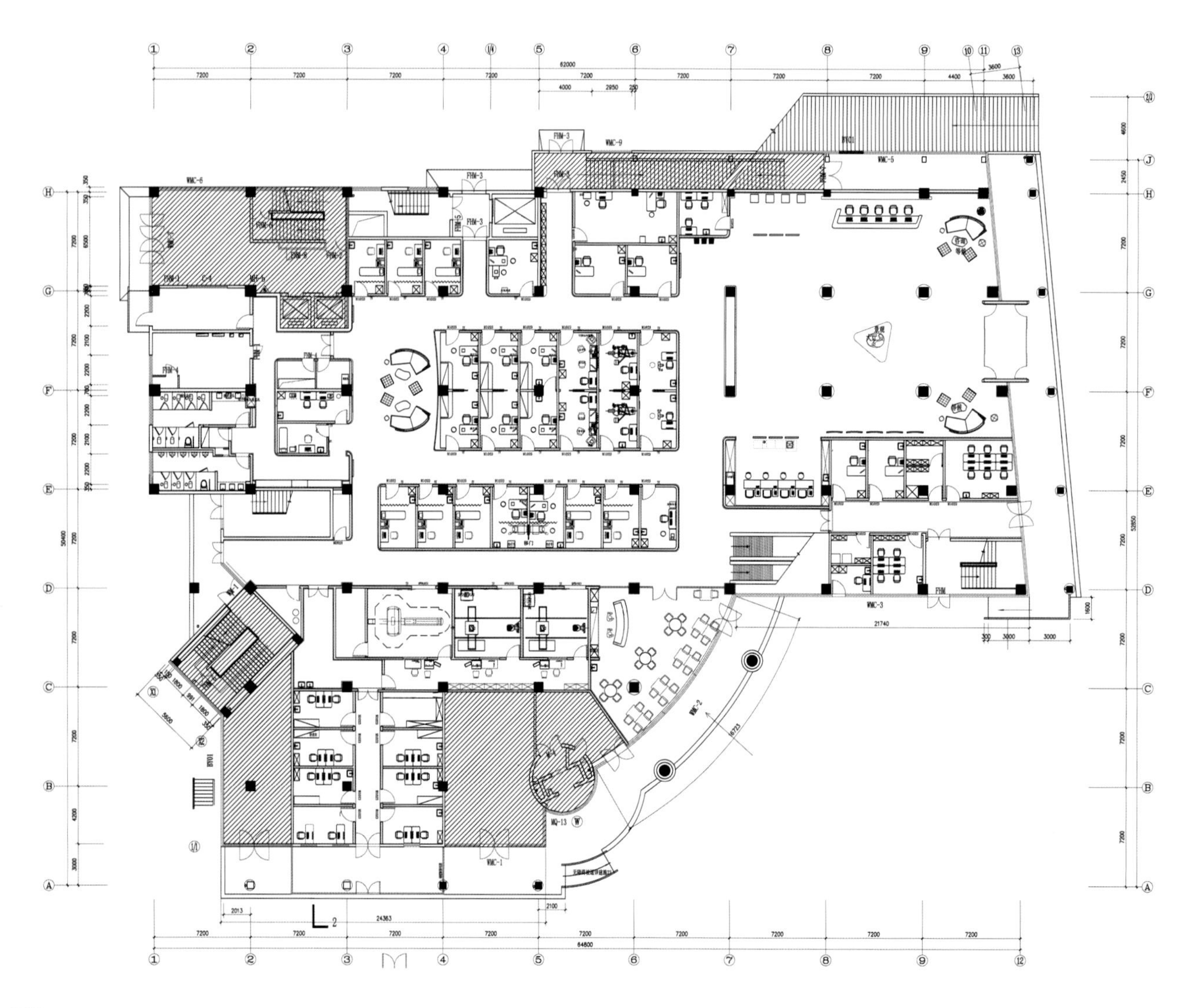

南通大
附属医院
检中心
STATION

STATION

STATION

体检中心国际部
International medical center
Information

“扬子江边第一珠，据川望海抱吴都。玉兰风度生清境，金菊精神浸画图。濠水秋光千古秀，文峰气象百年殊。”手法的延续与提炼源于意念，设计说明的构思原则为室内方案的形成提供了基础，各种元素与现代概念无形象意念的表达，是此项目设计的出发点。

Inspired by the elements of the Yangtze River, magnolia, golden aster and the Hao Water, the hospital design is the extension and refinement of thoughts. The design principles interpreted in this case has laid the foundation of the interior strategy. It is the ideological expression of all elements and modern conceptions that contributes the design orientation.

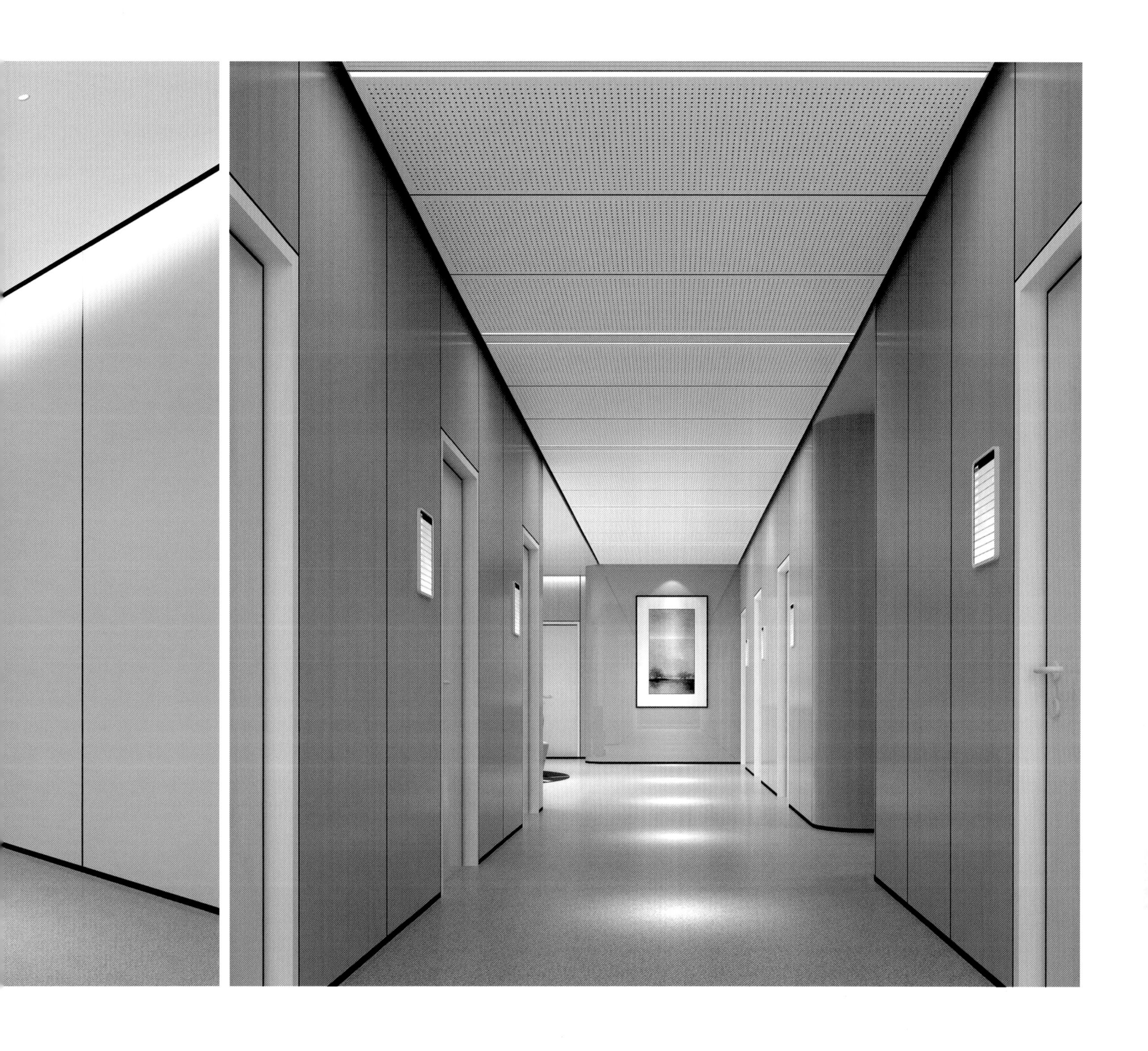

妇科康复

图书在版编目（CIP）数据

中国人文医疗空间开拓者：亚明大型医院室内设计 / 孙亚明编著；孙哲译. — 沈阳：辽宁科学技术出版社，2017.4
ISBN 978-7-5591-0119-8

Ⅰ. ①中… Ⅱ. ①孙… ②孙… Ⅲ. ①医院－室内装饰设计－作品集－中国 Ⅳ. ① TU246.1

中国版本图书馆 CIP 数据核字（2017）第 052880 号

出版发行：辽宁科学技术出版社
（地址：沈阳市和平区十一纬路 25 号 邮编：110003）
印 刷 者：深圳市雅仕达印务有限公司
经 销 者：各地新华书店
幅面尺寸：230mm×300mm
印　　张：37
插　　页：4
字　　数：200 千字
出版时间：2017 年 4 月第 1 版
印刷时间：2017 年 4 月第 1 次印刷
责任编辑：杜丙旭　孙　阳
封面设计：周　洁
版式设计：周　洁
责任校对：周　文

书　　号：ISBN 978-7-5591-0119-8
定　　价：328.00 元

联系电话：024-23280367
邮购热线：024-23284502
http://www.lnkj.com.cn

更多项目信息，请微信扫一扫